0
10
101

TOO TECH TO FAIL

THE DREAM, THE REALITY, & YOUR MUM'S ADVICE

written with love by
Gerard Salvador López

Cover: Gabriela Fariña
Interior book design: Sandra Herrera
Copyeditor: Mallory Miles
Proofreader: Vidya Vijayan

ISBN-10: 2960221400
ISBN-13: 978-2960221404 (AFNIL)
EAN: 9782960221404

To my parents and my sisters,
for always being there.

CONTENTS

CONTENTS

INITIAL WORDS

A. You bought a book. How cool is that?
i. Look at what you're holding

Let's analyze what just happened and get amazed by it. You, or the person who kindly bought this book for you, found it on Amazon, bought it online, and shipped it to a specific destination.

Nothing too crazy and exciting…*right?*

Simplifying of the process a lot, let's take a look at what the hell just happened.

You connected your computer through a network of computers, routers, and switches, called the internet, to a server that is probably located in another country or state.[1] Through different complex communication protocols, you exchanged information with this server and decided to place an order for one specific book: this one. After logging yourself onto the platform, you provided your local Amazon entity with some private information from a third financial party that both you and Amazon trust (VISA, Maestro, Mastercard, PayPal, etc.). Then, a new encrypted connection was made with another server (probably located in another country) that told both of you through a secure system like SWIFT if this order could be placed. The financial institution gave the OK, and your order was placed. The credit/debit card company became the intermediary to wire this bank transfer from your account at your commercial bank (probably located in your country of residence) to Amazon's account in your state. Then, Amazon moved this money around the world to pay for the printing of the book, my royalties, shareholders, taxes, etc. And so on and so forth.[2]

Up to this point, almost nothing has happened—just a few ones and zeros added to some servers spread around the world.

You placed the order through an interface from Amazon, but the company that publishes this book is CreateSpace, an Amazon subsidiary.[3] CreateSpace produces books on demand. You might think that your book was shipped from a stock, but it was actually printed a few hours after your "click." Your "click" just added a new item in the production line of a factory near you to which CreateSpace outsources work. Of course, the paper likely came

---

[1] Amazon servers are found in 12 different geographical locations.
[2] (Brown, 2013)
[3] In 2005, BookSurge and CustomFlix were both acquired by Amazon.com. CustomFlix's name was changed to CreateSpace in 2007. In October 2009, in a bid to promote harmony within the businesses, the CreateSpace and BookSurge brands were merged under the CreateSpace name.

from the forests of China or the United States (they have half of the world's market share in pulp and paper), and the printing technology and ink likely came from Japan or Germany.

Finally, your newly-made book was shipped through Amazon's local warehouses to your address.

If you think about it, your secure and straightforward purchase involved work from 10+ companies (including contractors and subcontractors) with actors coming from 10+ countries and under the legal framework and bilateral agreements of at least four different legislative bodies.

Now, multiply this purchase by the sum of all the ones you do online and offline. We are living in an incredible and exciting system full of interconnected dependencies. We are already living in a global city.

Welcome to globalization!

Cool, right?

ii. Why I spent my free time and energy writing this book

"Gerry, didn't you have anything better to do?" That's indeed a good question.

First, let me introduce myself. I am an engineer working for a worldwide industrial company. I am the typical guy in love with all new tech stuff and thrilled about the impact that new technology is having on our lives. I was so excited by the changes I had seen that I was spending crazy amounts of time reading articles and books about industry 4.0, big data, digitalization, NFC, bioprinting, industrial revolutions, 3D printers, Tesla, cloud computing, augmented reality, robotics, driverless vehicles, blockchain, Bitcoin, blah blah blah. Everything cool, indeed. But the more I read, the more I had the feeling that all these topics were being treated in isolation. We were forgetting the bigger picture and the real consequences of them on a macro scale. For example, genetic engineering will not only increase our life expectancy, it will also change our idea of life and death.

I wrote this book to try to present the enormous technological changes and the associated risks that we are going to face and to prepare society for them. This book is intended to be simple, factual, and straightforward. I've also tried to stay entertaining and funny, so you can enjoy the ride. Hopefully, you'll learn something new that you will be able to use to impress someone at dinner. Let's see what this new reality has in store for us.

Ready to take off?

B. Why is everyone so excited about change?
i. It's all about dreams

Once upon a time, three good friends were running a production company. Each of them had a dream, and they were working hard for it. The first one, the COO (Chief Operations Officer), wanted to mass-produce products of the best quality while keeping them as cheap as possible. The second one, the CMO (Chief Marketing Officer), wanted to be able to reach all potential customers with absolute customization of the product. The third one, the CEO (Chief Executive Officer), wanted to have a global vision while acting locally in each community. They all found the answers to fulfill their dreams in technology and, therefore, lived happily ever after. The end.

Cute story, right? Let's forget dreamland and be clear. Those dreams are as old as capitalism. By definition, expansion and growth are the ways to move forward in classical capitalist systems. And almost all of us (including China) are living in capitalist systems.

The words COO, CMO, and CEO sound fancy, but at the end of the day, all these people have the same dream. They want to grow and expand. It is even more important to realize that this dream has stayed alive through periods of economic expansion and contraction. It has nothing to do with economic cycles, and the industrial revolution that we are facing will not change our dreams of economic growth. However, it will shape the world faster than any other revolution has and will enable more progress toward the goals described by our three new friends.

ii. Dreams create revolutions

In recent years, the pace of innovation has been accelerating. Many things are changing, and it looks like we are on the cusp on a new industrial revolution that is reshaping our world. To begin with, we need a name for the revolution. Many technology buzzwords are floating around, but one has become the universal standard: Industry 4.0. If you are a bit more geeky, you might have also heard of the "Third Industrial Revolution." Depending on what you consider a revolution, you might agree that you are on the third or fourth step. Let's take a look at what these names stand for.

The term "Industrie 4.0"[4] comes from a German government project designed to promote the computerization of manufacturing. The narrative story explains that:

- The <u>first</u> industrial revolution mobilized the mechanization of production using water and steam power.

- The <u>second</u> industrial revolution introduced mass production to the markets with the help of electric power.

- The <u>third</u> industrial revolution further automated production lines by combining digital capabilities with production lines using electronics and IT.

- The <u>fourth</u> revolution[5] brings a merger of the virtual and real worlds in production plants.

---

[4] Yes, it's German. What a surprise. HA = D
[5] Or "Industry 4.0" (because it looks cooler).

The term Industry 4.0 was first used in 2011 at the Hanover Fair.[6] In October 2012, the Working Group on Industry 4.0 presented a set of implementation recommendations to the German federal government. The Industry 4.0 workgroup members are recognized as the founding fathers and the driving force behind Industry 4.0. Therefore, this nomenclature is part of the strategy of the German government. It makes strategic sense for the country because industry know-how is one of Germany's critical assets.[7] [8] [9] Table 01 describes each industrial revolution and its leading technologies.

	INDUSTRY 1.0	INDUSTRY 2.0	INDUSTRY 3.0	INDUSTRY 4.0
WHEN?	End of 18th century	Beginning of 20th century	End of 20th century	Beginning of 21st century
WHAT?	First steam-powered factories	Mass production thanks to electricity	More automatization with advanced electronics	Industry merging real and virtual worlds

TABLE 01: TIMELINE TOWARD INDUSTRY 4.0[10]

Besides this German terminology, you might have heard of the "Third Industrial Revolution." Its founder and most famous advocate is Jeremy Rifkin. In his definition, he states that revolutions are driven by major changes in energy, transportation, and communication. Those changes drive exponential growth and create a new economic paradigm that

---

[6] This fair brings together all the leading industrial companies in Germany to discuss the future of the industrial world.
[7] (IK4Tekniker, 2016)
[8] (Automation.com, 2014)
[9] (Deutschland, 2014)
[10] (Deloitte, 2015)

changes the way we live. Rifkin believes that we are currently entering the third industrial revolution with major changes in:

Energy. There is a shift in people's mindset about the way they produce and consume energy. The industry is moving toward more renewable sources of energy, and the consumer is moving toward self-production. It's becoming cheaper and easier to "go green." Moreover, we are starting to use big data to monitor our consumption patterns, which increases the efficiency of not only our homes but also the whole grid. Greater efficiency translates into cost reductions and a lower production need. The idea is to transform our buildings into renewable micropower plants with storage capabilities.[11] [12]

Communication. Internet technology can transform the power grid of every continent into an interconnected energy network that acts just like the internet. Millions of buildings generating small amounts of renewable energy will be able to sell their surplus, bringing energy back to the grid and sharing it with their neighbors. Moreover, the instant internet allows us to know what is happening all around the world in real time, which will make it easy to shift production in any industry.

Transportation. Transportation vehicles will transition to electric plug-in and fuel cell vehicles that can buy and sell green electricity on a smart and interactive power grid. They will be autonomous and will, therefore, reduce the congestion on the roads and decrease pollution in our cities.[13] [14] table 02 summarizes the three revolutions.

---

[11] (Gartner, 2015)
[12] (Salvador Lopez, 2016)
[13] (Rifkin, 2017)
[14] (Rifkin, 2015)

	1ST INDUSTRIAL REVOLUTION	2ND INDUSTRIAL REVOLUTION	3RD INDUSTRIAL REVOLUTION
ENERGY	Coal	Oil	Renewable energy & Smart Grid internet
COMMUNICATION	Steam Printing	Telephone	Communications & computing internet
TRANSPORTATION	Steam Locomotive	Internal Combustion Engine	Electric & autonomous transportation and logistics internet

TABLE 02: TIMELINE TOWARD THE THIRD INDUSTRIAL REVOLUTION

As we saw, there are two popular ways of naming and understanding this new revolution. But who cares? The important thing is to realize that we are at the front of a wild, global revolution that will completely change the way we understand the economy, society and life. There will be shining stars and people left behind. However, it will be our task as a society to make sure that the profits of this new revolution are shared globally.

For the rest of the book, I will use the words "new revolution" or "industrial revolution" to refer to this new paradigm.

iii. Revolutions create new realities

Our reality is changing. In a nutshell, the world is becoming digital. Every day, around 2.5 quintillion[15] bytes of data are created. As much as 90% of the data in the world today has been created in the last two years.[16] Sensors and connectivity between devices are the key. Gartner[17] estimated that there were 8.4 billion connected devices in use worldwide in 2017. That's up 31% from 2016, and will likely reach 20.4 billion devices by 2020. Total spending on Internet of Things hardware and services will surpass $2 trillion in 2018 and $3 trillion in 2020.[18] Those things are everywhere, and they do everything you can imagine. They generate data (lots of it) that can change our lives. Like it or not, our lives, our homes, and our planet will be more and more monitored in the next few years. Whether this surveillance benefits us or threatens us is up to us. We will enjoy all the power that technology has to offer, but we will be confronted with some ethical challenges too. Welcome to the new connected world.

Last but not least, I want to be clear about all the wizards out there predicting the future with their crystal balls. Nobody (exactly nobody) knows where we are heading. There are too many unknown and uncontrollable variables. Still, every publication that talks about the future is (or at least should be) based on current data, trends, and assumptions. This book is no exception. I tried to pull information from the most reliable resources available. I packed each page with references on purpose, so you can fact-check everything you read. I am not asking you to believe me blindly. Please always remain critical of everything you read, not only in this book, but also elsewhere in your daily life.

---

[15] One quintillion is a 1 followed by 18 zeros. This amount of data would fill 10 million Blu-ray discs. If we stacked those disks on top of one another, we would have a tower the height of four Eiffel Towers (Vcloud News, 2015).
[16] (IBM, 2018)
[17] Gartner is a global research and advisory firm providing insights and trend analysis.
[18] (Gartner, 2017)

"
I am one step away from being rich.
All I need now is money.
"

- Anonymous.

Grab your pumpkin caramel macchiato
with soy milk and vanilla flavor,
and let the show begin ;)

PART
ONE

A new beginning?

It is incredible to see how many great technologies are emerging in all industries, from biology to sociology. Life is becoming more comfortable, and we will be able to make more of our dreams possible. All these changes will affect the way we understand society as a whole. Any innovation has to be seen from a broader perspective. For example, self-driving cars are a reality that will not only change mobility but also change the way we understand distances. We might not mind sleeping in a car while it drives us to a business meeting that is located in another country. For the vast majority of people, the impacts of these new technologies are broader than the specific functionality they were designed for.

A. Will you have everything **you** want?

When we think about the rise of machines in this new industrial revolution, two images might come into our minds: a fully automated world as seen in Minority Report or Wall-E or the end of civilization due to uncontrollable robots as seen in The Terminator or The Matrix. Which one of these worlds will we live in? Will automation fulfill all our dreams and needs? Will we have everything we want?

i. Agriculture and cattle
Let's start from the beginning.

We all eat. And we have been eating since the beginning of time. Knowing that, you might expect us to have mastered agriculture centuries ago. The reality is that we are still in a phase of mind-blowing innovation. The use of closely-monitored operations in farms nowadays has turned them into factories. Technologies such as moisture sensors, automated irrigation systems, hydroponics, and artificial intelligence are already a common element in this new smarter agriculture. These innovations would not have been possible without the big leap that genome editing has brought about. Breeding in agronomy, horticulture, animal science, and aquatics has become more precise as we have developed a deeper understanding of the DNA sequences of crops and animals. Genome editing is projected to lower costs and boost yield, thereby increasing farmers' profits and regulating market prices. Breeding can also be a big help in developing sustainable farming techniques, especially now that the Food and Agriculture Organization (FAO) has released a report stating that by 2050, agricultural production must have grown by 70% to meet the demand of the forecasted 9.7 billion people who will be living on earth at that time. Data shows that we are not only capable of feeding this large population; we're capable of feeding them well.[19] [20] In 1900, around 41% of America's labor force worked on a farm; now that percentage is less than 2%.[21] This means that productivity (output per farmer) has increased exponentially. So, where are we heading? What is going on, and what new technologies are under development to increase that productivity even more?

The eruption of smart farming

Producing enough food to feed the world's population without harming the environment is probably the biggest challenge of our generation, and the farmers' skepticism toward change has made the problem even more difficult. However, who could blame them for their caution when they could lose everything by making the wrong decision? Every farmer needs to monitor the weather, soil health, water supply, and nutrients, as well as ward off weed competition, pests, and diseases to ensure optimal yield and profit. Smart farming will make their production system more cost-effective and less laborious; they will have technologies like GPS on tractors and mobile machines, microbe-based products (e.g., pesticides that enhance nutrients in soil), high-density soil sampling, and yield maps to do the hard work for them. Drones, satellites, and other airborne instruments can also gather information about plant health and soil status. These innovations have already resulted in a shift in the business models of big firms like Monsanto, Dow, DuPont, Bayer, and Syngenta.[22] Agricultural

---

[19] FAO is the Food and Agriculture Organization of the United Nations. Its vision is to achieve food security for all to make sure people have regular access to adequate high-quality food to lead active, healthy lives. They have great projects in developing countries funded by the United Nations and private parties.

[20] (The Economist, 2016)

[21] (OECD, 2011)

[22] Globalization is convincing industry giants to merge to remain competitive and able to offer complete solutions to clients. Listed below are some of the most famous M&As in the market:

- Dow and DuPont, two American giants, are planning to merge to create DowDuPont. The resulting company will have an estimated market value of $130 billion at the announcement and will divide its business into three independent companies (Agriculture Company, Material Science Company, and Specialty Products Company).

companies have started investing in software platforms that will serve as a farm's management system, programmed to make recommendations to farmers based on the data collected, the farm's history, crop profiles, and weather forecasts. Some have gone further and started developing robots as well. You can think of farming as a matrix like the ones you studied in algebra when you were younger. A farmer must play with a significant number of variables to optimize his yield and maximize his profit. Smart farming has two applications for the farmers and the companies that supply innovative agricultural solutions. The first is to measure the variables going into the matrix as accurately and cost-effectively as possible. The second is to suggest procedural changes based on all the data collected. Technology is taking over agriculture and bringing this science to a new level, and anyone can engage with this industry without much background knowledge. Agricultural firms are into technology platforms that increase productivity while also decreasing the workforce size, thanks to robotized machinery. Take a look: apart from selling motors, seeds, or chemicals, the big agricultural players have developed software platforms that can adequately manage your farm. What makes this software advanced is its ability to identify the behavior of crop strains as well as local weather forecasts. By combining this information, the software can provide appropriate recommendations to farmers, making their crops more productive and profitable. The functionalities of farming management software are increased through the use of sensors, which are becoming more and more affordable. Sensors can help you analyze many parameters of the land, like the bearing

---

- Monsanto, another prominent American firm, is the subject of a takeover bid by Bayer, a German multinational company. Bayer offered 66 billion USD, and Monsanto shareholders approved it.
- Syngenta, a Swiss company, is being bid by ChemChina, a Chinese one, for $43 billion. It is China's most ambitious takeover deal in history.

The key to all these moves is to be able to offer integrated solutions for the firms' clients and to take advantage of the important synergy opportunities from the economies of scale and scope. The rise of new business models requires collaboration between companies and, if needed, M&A solutions. At the time of the writing of this book, these M&As are still subject to regulatory approvals.

capacity and pH of the soil. They help you identify the amount of pesticide for a particular soil. Some businessmen have also profited by going solo and forming networks or cooperatives instead, in which the group purchases the licenses for a software that everyone can use.

The future of aquaculture

The current state of the world's oceans is somewhere between poor and very poor, and that's being optimistic. The degeneration of the oceans is happening much faster than scientific studies (or the media) have predicted. What's worse, the conditions currently distressing the marine environment are similar to those that were associated with all major extinctions in Earth's history: an increase of both hypoxia (low oxygen) and anoxia (lack of oxygen that creates "dead zones") in the oceans, warming of ocean temperatures, and ocean acidification. The combination of these factors will inevitably cause a mass marine extinction, and the damage is already evident.[23][24][25]

Despite the degeneration of the oceans, the production of fish for human consumption has risen from 20 million tons in 1950 to 170 million tons in 2014. However, the volume of caught fish has been stuck at around 90 million tons since 1990. What has filled the gap? Farmed fish.[26] The amount of farmed fish surpassed that of beef by a million tons in 2013 and will exceed the volume of caught fish in 2020.

To lessen pressure on the oceans, aquaculturists are always looking to create artificial ecosystems that are as real as possible for the fish. This is where the magic of inland fish farming comes in. One of the major problems of inland farming is the greenhouse gas emissions produced by the aquaculture process. But this is going to end.

---

[23] (Donovan, 2011)
[24] (International Programme on the State of the Ocean, 2013)
[25] (Than, 2016)
[26] Fish farming or pisciculture involves raising fish commercially in tanks or enclosures such as fish ponds, usually for food.

Dr. Zohar and his team are aquaculture gurus. His ecosystem continually recycles the same supply of brine,[27] purified by three sets of bacteria. One set turns ammonia excreted by the fish into nitrate ions. A second converts these ions into nitrogen (a harmless gas that makes up 78% of the air) and water. A third works on the solid waste filtered from the water, transforming it into methane, which provides part of the power that keeps the whole operation running. The best part is that this system is closed and can be set up anywhere. It doesn't cause pollution, and is disease-free. What makes this operation amazing is its ability to grow species of fish which could previously only be bred in the wild. Such rare (and expensive) fish species can now readily be made available for consumption all over the world.

Furthermore, today's fish farmers are using transgenic techniques.[28] In other words, they mix a particular gene with an existing genome that results in the rapid growth of a species (or bestows other superpowers such as better taste or higher survival rate). Gone are the days when fishermen had to comb through fishing nets, sorting fish by their size. It is now possible to monitor growth and quickly identify the appropriate time to harvest fish. You don't even have to wait for the seasons of natural spawning. For example, salmon, which is naturally in season in spring and summer, can now proliferate all year round.[29] Size isn't the only thing you can hack with transgenes. Researchers from salmon breed in Norway still study DNA but focus on how DNA could resist disease and increase survival by tracking SNPs (single-nucleotide polymorphisms). Using this method, researchers were able to produce salmon resistant to sea lice and pancreas disease, a classic viral disease that affects farmed salmon.

Although transgenes are still controversial, their use is expanding as the demand for fish continues to grow. Technology is here to

---

[27] Brine is a high-concentration solution of salt in water.

[28] Transgenic refers to an organism whose genome has been altered by the transfer of a gene or genes from another species or breed. In research, transgenic organisms are used to help determine the function of the inserted gene, while in industry, they are used to produce a desired substance.

[29] (AquaBounty, 2017)

help provide us good fish all year long. Also, guess what, thanks to independent ecosystems and genetic engineering, we are managing to meet the increasing demand for fish necessitated by the ever-booming global population.[30]

Genetically Modified Organisms or GMO

Decades have passed since farmers first started using genetically modified organisms (GMOs), but some people are still unconvinced that GMOs are safe for human consumption and for the environment.

Admittedly, transgenic modification is a haphazard technique since the result of moving genes from one species to another is unpredictable. But genome editing is different. Rather than shuffling entire genes between species, genome editing only adds, subtracts, or substitutes a single nucleotide in the genetic code. Experts are positive that genome editing can change the consumers' perspective about GMOs because, aside from being precise, the process mimics natural and random mutations.

Synthetic biology, which includes transgenic modification and genome editing, is a science that involves the collaboration of molecular biology, biotechnology, genetic engineering, systems biology, biophysics, and computer engineering. Considered the leader of the next industrial revolution, synthetic biology is rapidly evolving and is basically present in every aspect of human life. Food, clothing, and medicine are all produced using synthetic biology.

Synthetic biology is rooted in genetic engineering and creates a standardized and automated method of modifying an organism's DNA code to create a new kind of organism or to improve certain characteristics of an existing organism. Due to the establishment of DNA databases, DNA can now be put together like LEGO bricks to create a new organism, correct a defect, or make an organism resistant to a disease. This technology has huge potential to impact food production. It might become the

---

[30] (The Economist, 2016)

solution to producing more nutritious crop varieties despite the effects of climate change and the limited supply of water, energy, and land. In addition, the cost of DNA sequencing is now much cheaper than it used to be. So, if combined with automation, the result would be a more refined methodology for genetic engineering.

However, the great progress being made in synthetic biology has sparked concerns from the public. To some, synthetic biology seems like an unnatural process, a manipulation of life often referred to as "playing God." The fact is that synthetic biology is just another type of innovation. Gene modification techniques and selective breeding of plants and animals have been used by humans since the birth of agriculture. People will eventually get accustomed to synthetic biology if it becomes a new normal standard. Consumers might one day start to appreciate the benefits of synthetic biology when it is incorporated into other fields, such as medicine, which will save their lives, or bio-fabrication of clothes and tissues, which will prevent animal abuse.[31]

Final thoughts

Indeed, the future in front of us seems bright when we speak about agriculture. This industrial revolution will not only change the efficiencies of the land, it will also have significant implications in food security and the overall concept of farming. We will need even fewer people to work the land, and the type of jobs these workers have will change dramatically; programmers and genetic engineers will be in greater demand in the agricultural industry. Supplying enough food for humanity, even when our population hits 10 billion, can easily be ensured by the technological advances described. Businesses are putting their efforts toward making these solutions more global and efficient. If we want to leave the Earth to live on other planets some day, we will need these advances to succeed in our space odyssey. One last thought: you might sometimes hear the claim that there are too

---

[31] (McKinsey & Company, 2016)

many people and that we won't be able to feed everyone in the world. This is just false. The world already produces more than one and a half times of food necessary to feed everyone on the planet, which is enough to feed 10 billion people. The problem is not our ability to produce food. Hunger is a social problem; people who earn less than $2 a day can't afford to buy food, no matter how abundant it is.[32] [33]

Box1: Improving animal welfare

People often assume that technology will bring about a de-humanization of agriculture, since machines will be heartless in their treatment of animals. But the opposite story might be the case. It is true that current practices skew toward giving preventive antibiotics and shortening the lifespan of farm animals. However, these trends are hopefully going to change for three main reasons.

The first reason is the power of worldwide social media. The unfair treatment of animals in the agricultural industry is nothing new. However, in current times, unfair practices are more likely to be documented and discussed. The results are numerous: hundreds of animal welfare documentaries, an increasing number of vegetarians and vegans, thousands of articles about farming practices, new labels for ecological products, stricter regulations, etc. All these changes have arrived merely because of public awareness, the active role of international NGOs, political pressure, and dynamic environmental lobby groups.

The second reason is quality. If a farmer with cattle wants to sell good meat, it is essential that the cows live a healthy life and do not have diseases. This is where technology plays a role. Cattle are now implanted with sensors that can monitor the acidity of their stomach and other digestive problems. SmartBell, based in Cambridge, UK, has developed sensors that can be hung around the cattle's neck to monitor their movement and health. This type of sensor can detect problems at an early stage before they get

---

[32] (Nature, 2012)
[33] (Huffington Post, 2014)

worse. Moreover, slaughtering techniques have completely changed and are heavily regulated, ensuring fair treatment in the last minutes of an animal's life.

The third reason is that genome editing has the potential to increase the welfare of animals. For example, the bio-engineering firm Recombinetics was successful in creating a hornless Holstein cattle through genome engineering. Holsteins are an excellent source of milk, but their horns are difficult to handle and are very dangerous to farmers and other animals. To address this problem, the animals are typically dehorned when they are young in a process that is painful. Therefore, breeding hornless Holsteins notably reduces their pain.[34] Another example comes from the researcher Randall Prather from the University of Missouri, who was able to breed pigs that could resist porcine reproductive and respiratory syndrome (PRRS), a viral disease that has been affecting pigs for decades and costing farmers large amounts of money (more than $600 million a year). Steve Kemp and his colleagues at International Livestock Research Institute in Nairobi were able to make cattle resistant to sleeping sickness disease, which has been a massive killer of livestock. Bruce Whitelaw at the Roslin Institute in Scotland made pigs immune to African swine fever by altering some genes. All these gene editing examples have been helpful to farmers and, of course, to the animals themselves, giving them a healthier life. These alterations were able to improve productivity and welfare.[35] The difference in what we call natural and what we call artificial is becoming blurry. The question of whether or not it is a good idea to play with the laws of nature is, of course, a philosophical question for another thoughtful book.

Box 2: Hydroponic solutions and LED technology

It is generally assumed that vegetables and other crops should be grown in large sunny areas. However, the future is here to tell us that this is no longer a requirement. New hydroponic

---

[34] For the curious ones, the process of making hornless Holsteins involves deleting a sequence of 10 nucleotides and replacing it with 212 others.
[35] (The Economist, 2016)

solutions, combined with LED technology, make vegetable gardens possible anywhere.

Hydroponics is a method of growing plants without soil, using mineral nutrient solutions in a water solvent. Terrestrial plants may be raised with only their roots exposed to the mineral solution, or the roots may be supported by an inert medium, such as perlite or gravel. A tunnel in Clapham, South London, that used to serve as a World War II bomb shelter is now used for growing underground plants. Beautiful metaphor, right? Sounds impossible, but the tunnel is currently growing 20 types of different salad plants, mainly delivered to chefs and salad shops in the city. You may have already eaten salads that have never felt the sun without even knowing it.

Underground farming resembles indoor hydroponic methods. The big difference is that, instead of redirecting sunlight to the crops, the light is provided by efficient and long-lasting light emitting diodes (LEDs). These LEDs are tuned precisely to the plant to contribute best to its photosynthesis. Here's the key: each type of plant is different and needs a specific amount and wavelength of light. In other words, not all the sunlight is useful for the plant. Knowing the particular needs of the plants makes production more efficient and reduces the growing cycles, making this hydroponic business more profitable. And don't forget, the new hydroponic method doesn't always have to be underground. In some places, crops are grown vertically in buildings like abandoned properties, old factories, and warehouses. All this means that farming could become possible in urban areas. Hectares of soil won't be necessary anymore. With this advancement, you can venture into agri-business even when you're in the city.[36]

Box 3: The invasion of the sky

There is also a noticeable change in the air. The early rise of agricultural drones has helped farmers examine crop health by

---

[36] (The Economist, 2016)

flying cameras along agrarian surfaces. This procedure requires less people yet provides an accurate examination of crops. Agricultural drones exhibit excellent performance in the farming industry, especially quadcopters that are smaller in size than regular drones. Quadcopters might become the standard in the industry even if they now have a limited range and endurance over farm surfaces. There are many models to choose from: Google AgDrone built by HoneyComb, Lancaster 5 by PrecisionHawk, or DelAir Tech based in Toulouse, France, to name a few. They are all fancy and impressive projects.

And drones aren't the only newcomers in the sky. Low-height satellites have become popular, offering a broader field of vision and automatic, periodic examination. The Planet Lab based in San Francisco was able to program satellites to take pictures of a farm's surface at least once a week. They were able to identify problems with their detailed examination, which helped farmers to quickly address problems with their crops, resulting in a better harvest and bigger profit margin. Farmers opt to spend more on these technologies because they are worth it.[37] [38]

---

[37] (Forbes, 2013)
[38] (Planet, 2018)

ii. Manufacturing and non-financial services
To begin with

We live in a capitalist world. Every new good produced and service provided is counted in a number called Gross Domestic Product (GDP). The higher the GDP is, the more successful and powerful the country is. Trade is the key to increasing GDP because it gives the nations the ability to export what they are best at and import what they lack. Even if it doesn't sound like common sense, trade is not a zero-sum game. Both parties can win. The core of any industrial revolution is an increase in production levels due to the efficiencies brought about by new technologies and expansions to new markets. The current revolution is no exception. The whole manufacturing and service industry is being disrupted and rebuilt by new technologies. The untapped potential is still under construction, but the future looks promising. What is the new mindset we are heading toward? What are the trends of the sector?

A wedding between the digital and the physical world

The whole production concept is going to change radically—not only the processes but also the way we think about consumption. All the products that we consume will be tailored to satisfy our own needs, personality, and tastes. Companies will be able to offer high levels of personalization while keeping up mass production. To achieve that goal, the production processes (and even the factories themselves) are becoming more and more digital. In the words of Joe Kaeser (CEO of Siemens), "We merge the real world, which is hardware, with the virtual world, which is simulation software."

The German conglomerate is not an exception in the market. The whole industrial sector is moving toward digitalization. Today, manufacturing lines and processes are designed even before the factory plant is built. Computers and software allow you to create a real-time copy of the manufacturing processes in a virtual world that optimizes engineering, processing quality, uptime, and load time. When all this analysis is done, a refined production process can be copied back into the real world. Production plants go from the virtual world, where everything is planned and optimized, to the real world, where everything is constructed and produced. When processes are simulated in manufacturing, companies can go from destructive to non-destructive testing. For example, Boeing simulates the whole development and engineering of their new airplanes. Then, they simulate whether the airplane that they are going to produce can fly or not. All this happens long before the production line is even settled![39] How cool is that?

But let me tell you an exciting secret: what I just described is part of the present. Technology has much more to offer in the upcoming years. It is no surprise that General Electric has the ambition of becoming one of the top 10 software companies by 2020.[40] Companies building power plants and gas turbines want to hire more programmers than software companies do. Predix (GE), MindSphere (Siemens), and Uniformance (Honeywell) are all platforms that allow this digitalization of the industry. And the rest of the market is following. It seems that there is a new motto for industrial players: "Go digital or go home!"

Now let's take a look into the future. It's a rainy day in autumn 2030. The head of a team in a random manufacturing company has been assigned to create a robotic flying chair. This robotic chair is expected to be able to fly along the corridors of a new central station that will be inaugurated. It has to be operated through voice commands, internet commands, and joysticks. It is expected to be capable of traveling 20 kilometers and then returning to its point of origin, if necessary, to charge itself. Ah, I forgot. The robotic flying chair should be ready in 60 days, and nobody has ever created a chair like it. What will happen? Simple. Top engineers will be hired from all over the world to work on the project. Of course, hiring brilliant minds is still a must. But they are easy to find in global databases of freelancers. When the multicultural team is formed, they don't need an office. Each of

---

[39] (Kaser, 2016)
[40] (General Electric, 2015)

them is in his or her own home in front of a computer. The team leader will share views, the client's idea, and general information through a database that everyone in the team can access. Then, he will upload a file named "XtraScoot" which contains all specifications required. As soon as the file is activated, an AI program will automatically search all the databases of the different suppliers of the world for self-inflating tires, brake systems, possible designs, and any devices that would possibly make the XtraScoot a fantastic product. In just a few minutes, a list of potentially applicable components, complete specifications, availability, earliest delivery times, and prices will appear. It will become possible to assemble 3D interactive models. Of course, all that information will be available to the members of the team and securely stored in the cloud. Now, finally the humans play a role. The AI has provided a solution that is compliant with the request of the client, but the solution proposed is not yet aligned with what the customer really wants. To create something the client really wants, the humans have to read between the lines of the contract. The engineers give their ideas and opinions and work on each part according to their expertise. The mechanical, electrical, software, automation, and production planning will follow, always with the help of virtual assistants. The system itself works as a team: if one expert alters something from his area, everybody on the team can see the alteration. Some new features are added to make XtraScoot an outstanding robotic flying chair. As soon as the prototype design is finished, a corresponding virtual prototype of the production process is assembled. With the approval of the final client, the team of engineers will survey the market to find factories that have the required equipment to carry out the production process. They will merely rent some factories for a while (they only need 50 units). Finally, they will send the production information, and 3D printers and automatic processes will do the rest. After 60 days, the chair will be ready for the inauguration. Cool, right? And the coolest part is, the technologies described in our XtraScoot story are not as futuristic as they might sound. Many of them are already under development.

Conventional wisdom holds that the competitive advantage of any manufacturing process has three degrees of freedom: speed, quantity, and quality. You can try to excel at two of them but never at all three at the same time. For example, if you have the best quality product on the market and you produce large quantities of that product, you will not be able to have the fastest delivery time.

The new industrial revolution will completely reverse this statement. With the help of technology, even the most complex products will be successfully produced in short periods of time. The virtual world will change the way we understand manufacturing.[41]

The incredible power of automation

Humanity has always tried to do more with less. The definition of productivity has always been present in economic books to describe how well a society is doing compared to others. And the level of automation is intrinsically linked with the productivity of a country. Automation can reduce errors, improve quality and speed, and even reach outcomes beyond human capabilities. Productivity growth can boost economic growth and prosperity and help offset the impact of an aging workforce in many developed countries. It is expected that the global productivity growth from automation could rise by 0.8 to 1.4% annually. Just to have an idea of this change, the productivity growth from the steam engine was 0.3% (from 1850 to 1910), from early robotics was 0.4% (1993 to 2007), and from early IT platforms, 0.6% (1995 to 2005).[42] Many labor-intensive firms will be able to boost profit margins as they substitute costly workers for cheaper robots or intelligent software. And a range of entirely new companies and sectors will spring into existence.[43]

In addition to boosting productivity, automation will make it easier to personalize the products that are available on the

---

[41] (Siemens, 2014)
[42] (McKinsey & Company, 2017)
[43] (UBS, 2016)

market. Big data will be able to determine the type of products you want and offer them to you. If you remember the CMO from our story of the three friends who founded a production company, he "wanted to be able to reach all possible customers with absolute customization of the product." This will be possible because the cost of ordering one product or fifty products will be the same per new unit. That's the magic of automation![44] Let's imagine that 10 years from now, you want to buy a car (assuming that you will still want to own a car)! In the very near future, you will go to the website of your favorite car manufacturer, put your specs together, and send that order to them. A computer will check your credit history and your available funds. Then, your car will go straight to production, and the factory will build it customized for you with all your desired specifications. Finally, you will receive a car at your home. Not just any old car, but a car made for you! You will no longer need to wait six months or compromise at the dealership, where they may have hundreds of cars but not the car of your dreams.[45] In other words, when the smart factory is set up, it will be ready to produce personalized orders automatically while gathering data that will be used to reduce marginal costs and boost productivity. The main beneficiary will be the customer, who will be able to buy cheaper, unique, made-on-demand products.

Moving toward services, platform orientation, and freelancing

Traditional business models are moving toward services. They don't have a choice. Nowadays, Nike doesn't just sell shoes. They sell coaching services through apps to their clients, so the overall experience with the brand is enhanced.[46] This is just the beginning…

Services are getting commoditized. Just consider these trends seen in the younger generations:

---

[44] (Kaser, 2016)
[45] (Rifkin, 2017)
[46] (Nike, 2017)

They don't want to own a car anymore. They want to have services like Uber or Cambio.

They don't want to own CDs. They want to have instant access to any song that ever existed through services like Spotify or Google Play Music.

They don't want to go to a doctor once a year. They want to have a wearable monitor like the Fitbit watch or the Apple Watch that tracks their health 24/7.

Globalization and automation have made goods easy to replicate and always accessible. Companies will have to fight to provide the best customer experience through services to remain alive and relevant in the market. They will have to be able to change and adapt their business models to become more service and customer oriented.

Besides, the world's leading companies have realized that the future is *owned* by collaboration and collective intelligence rather than secrecy and closeness. The barriers to entering a market are so low that businesses don't have a choice and have to collaborate with each other. One of the biggest assets that Google and Apple have is their app market. They have created a platform on which developers and freelancers can make a living. The *platform companies* charge a fee out of the income that those apps generate. Their users and programmers bring the real value to this market, not the platform itself. Google Play and the Apple App Store have more than 3 million apps each, available to all Android and iOS users. On the business-to-business (B2B) side, something similar happens. Platforms like Predix® from General Electric allow independent developers to build their programs and enable them to sell their work to third parties. It seems that everyone is becoming more and more *open to outsiders* using their systems. This trend will move forward in the years to come.[47] [48]

---

[47] (Statista, 2018)

[48] (General Electric, 2018)

Last but not least, outsourcing is becoming easier. Today some firms are outsourcing almost everything, leaving only their core business *in-home*. Prestigious (and expensive!) consultancies have teams overseas to correct their slides. Famous telecom operators have their customer service in countries located far away from where the real customer is. Well-known authors have parts of their articles and magazines written and corrected by somebody else. At first, it might sound like exploitation, but this system is good for individuals too. You can hire a private English tutor living in the Philippines to improve your language skills for a small fee. You can forget about all the crazy tax bureaucracy for your small business by letting somebody else do the accounting. Freelancers and outsourcing companies are fast, flexible, available for short-term projects, easy to reach, and typically very cheap. Globalization and the internet (if you can even separate those two terms) are driving this new outsourcing attitude. Companies and individuals will have more free time to focus on their core competencies and/or their passions if some their work can be effectively outsourced to freelancing professionals.

Final thoughts

Indeed, the future in front of us seems bright when we speak about manufacturing. This industrial revolution will not only make production processes more efficient, but it will also change the way we understand uniqueness and personalization. The reduction of marginal costs to near zero will introduce the customer to a world with endless possibilities, and the power of platforms and outsourcing will make this world a smaller place to live in, allowing people to work on their competitive advantages. The speed of change coming from globalization is intrinsically linked with this industrial revolution. They are concepts that feed each other all the time. The more this revolution moves forward, the more globalization will be enhanced and vice versa. It is unstoppable. We are lucky to be living in times where we can witness all those changes and be part of them.

Box 1: Siemens and Drones

Today, countries and large industrial players typically use pipelines to supply whatever they need (usually oil, gas, heat, or water). This method has been regarded as the safest and most effective means of distributing substances for decades. Pipeline operators need to monitor the whole network to make sure that critical pipelines are still functioning well and are safe in general. To examine a pipeline's operation, aerial monitoring is conducted through the use of helicopters. However, flying over pipelines every two or four weeks for visual inspections can get expensive. Aside from cost, helicopter monitoring lacks equipment that can determine potential problems. To cut down on expenses and eventually lift up standards too, Siemens has developed a systematic inspection system that uses drones. Drones are equipped with 3D image cameras, color and near-infrared (NIR) sensors, and smart image analysis technology to monitor pipelines, large industrial facilities, and pylons in high-voltage electrical distribution systems. They are programmed to fly pre-determined routes over the surface. If an anomaly is detected, the data is sent to the pipeline operators to help them identify the problem and come up with possible solutions. What is even more interesting is that the drones can detect a problem that hasn't occurred yet (prevention is usually cheaper). But it doesn't stop there! This technology can also be applied to monitoring tall buildings and large industrial offices. Buildings consume a lot of heat energy, making it necessary to ensure that no heat escapes through poor insulation. With drones, areas of the building with insulation problems can be easily identified with infrared cameras, implying cost savings and improved energy efficiency.[49]

Box 2: Virtual reality to sell Volvo XC90s?

Virtual Reality (VR) is a technology that takes the user into a virtual world, and Augmented Reality (AR) is a technology that fuses the physical world with a digital world. They can influence people's

---

[49] (Siemens, 2016)

daily activities, from medical consultations to watching sports on Saturday nights. These two modern technologies, which are still in their early phases of development, are already expected to raise a multibillion-dollar industry within the next couple of years. VR/AR will pave the way to the widespread acceptance of head-mounted display devices. There are nine potential markets for this technology to be disruptive: education, healthcare, engineering, military, retail, real estate, live events, video entertainment, and video games. Among them, healthcare, real estate, and retail are projected to bear the most significant impact.

Volvo's VR campaign is an excellent example of the potential of virtual reality. The company produced the first virtual reality test drive for the upcoming Volvo XC90 though a VR application that you could download on Android and iOS. The app allowed potential customers to test the new car anywhere anytime and customize the characteristics that they wanted to have (i.e., the color of the exterior, sport or classic version, automatic or manual transmission). The result was that the first-edition XC90s sold out in less than two days with the app downloaded over 40,000 times. Of course, it is not only the fancy app that made this launch a success. But it is clear that the way we sell products is changing. Marketers are taking advantage of the innovations coming from the merger between the digital and the physical world to make the overall brand experience more powerful.[50] [51]

Box 3: 3D printing used by Vectoflow

3D printing, or additive manufacturing, is a technology that is being developed in this industrial revolution. Without the need for hard tooling, less time is required to build prototypes and final pieces. 3D printing, therefore, cuts turnaround times and lowers costs. Fields such as medicine and aerospace that account for low production volumes, high labor costs, long production

---

[50] (Goldman Sachs, 2016)
[51] (IOT for all, 2017)

cycles, and complex tooling can now be further developed thanks to this technology.

The aerospace company Vectoflow demonstrated the power of 3D printing with its Kiel probe. Speed is one of the most important and critical factors for any aircraft. If it is too slow, the airflow can stop abruptly, causing the plane to crash. If it is too fast, there will be too much stress on components, which can lead to failure. In normal planes, small probes measure speed based on the pressure of the airflow. However, the high forces, high temperatures, and unpredictable airflows make the manufacturing of these components very challenging. Vectoflow is specialized in the production of these probes. The company uses the additive manufacturing technology developed by EOS to produce probes in a single piece and smaller size, that have less air resistance. The flexibility in design, size, and material, coupled with fast production and delivery, has made 3D printing a popular choice in the aerospace engineering company. Additive manufacturing has shown that it can work efficiently even where maximum safety standards are a must.[52] [53]

---

[52] (EOS, 2018)
[53] (McKinsey & Company, 2014)

iii. Energy
To begin with

Energy is everything. Nothing works without it. Imagine how apocalyptic it would be to not have electricity for a day. No cars, no heat, no industry, no trade, no way to use money, no Amazon, no flights, no phone calls, no Netflix, no Facebook, no cat videos on YouTube... absolute chaos. In developed countries, we tend to forget how dependent we are on a stable supply of energy. We cannot imagine a world without it.

In 2015, the world's total primary energy supply was almost 160 million GWh, which is around 23 kWh per person per year.[54] [55] That's a lot of energy, but it differs depending on the development level of the country. As surprising as it might sound, more than three-quarters of the energy consumed in the world comes from fossil fuels, and only around 4% is renewable. Four percent! We are simply devouring our planet's resources, knowing that they are scarce and will one day run out. What's more, it is expected that the global energy consumption will rise 28% between 2015 and 2040.[56] [57] [58] We have an urgent problem that puts our survival and development at risk. The question is: how long will we be able to maintain this pace of energy

---

[54] (International Energy Agency, 2017)

[55] Primary energy is energy found in nature that has not been subjected to any human-engineered conversion or transformation process. It is energy contained in raw fuels and other forms of energy received as input to a system. Primary energy can be non-renewable or renewable. Primary energy sources are transformed in energy conversion processes to more convenient forms of energy that can be used directly by society, such as electrical energy, refined fuels, mechanical work, etc., with an inevitable loss of energy in the process. The use of primary energy ignores conversion efficiency: some forms of energy have poor conversion efficiency (e.g., thermal sources such as coal or gas that reach between 30 and 40%) whereas some have high conversion efficiency (e.g., hydroelectric plants that reach 90%).

[56] Fossil fuels are created by the decomposition of dead, buried organisms. It takes millions of years to convert the carbon in organic matter into fossil fuels like petroleum, coal, and natural gas.

[57] Renewable energy is derived from natural processes that are replenished at a higher rate than they are consumed. Solar, wind, geothermal, hydropower, bioenergy and tidal power are all sources of renewable energy.

[58] (International Energy Agency, 2017)

consumption? Is the renewable revolution capable of supplying adequate energy on a global scale? Can this industrial revolution become the answer to the energy crisis?

Renewable energy as an answer

Let's take a look at the world as it is today. At the time this book was written, the best data available was from 2015, and I am sure it will surprise you. table 03 describes the total primary energy supply in the world.

	WORLD'S PRIMARY ENERGY SUPPLY IN %	WORLD'S PRIMARY ENERGY SUPPLY IN MTOE	RENEWABLE
OIL	31.7	4,326	No
COAL	28.1	3,835	No
NATURAL GAS	21.6	2,948	No
BIOFUEL AND WASTE	9.7	1,324	Debatable
NUCLEAR	4.9	669	No
HYDRO	2.5	341	Yes
OTHER (INCLUDES GEOTHERMAL, SOLAR, WIND, HEAT, ETC.)	1.5	205	Mostly yes
TOTAL	100	13,647	

TABLE 03: FUEL SHARE OF TOTAL PRIMARY ENERGY SUPPLY IN 2015[59][60][61]

---

[59] This table includes the production of electricity.
[60] (International Energy Agency, 2017)
[61] Mtoe (Million Tons of Oil Equivalent) is an energy unit. To have an idea, 1 Mtoe is 11,630 GWh (Gigawatt hours).

The current data is a clear indication that we are far from having a sustainable way of feeding our energy system. The projections show that the world will continue to depend on fossil fuels until 2040, despite the projected boom in wind, solar, and geothermal energy production. Coal and natural gas will probably still dominate the global power supply because they are cheap, and we already have plants to process them. As in any projection, there is a certain degree of uncertainty. If the world leaders put up a united and aggressive front against climate change, the share of renewable energy in global electricity production may change between 17% and 31%. However, it will still be far from the 100% that is desirable.[62] [63] [64]

Let's take a look at the global rise in energy demand and its source. The growth rate may vary because of factors such as moderate economic growth and population growth, digitalization, economic shift to services, fall of demand in Europe and North America, greater efficiency in the energy sector, etc. However, there is a new concept that needs to be introduced: global energy intensity (GEI). The rate represents the amount of energy used to produce one extra unit of GDP. GEI is expected to decline by about 50% from 2013 to 2050, partly because electricity is estimated to cover 25% of total energy demand by 2050 (compared to the current 18%). Electrification will indeed be one of the main drivers of the changes that we can expect in the world. The foreseen rise in electrical production and consumption will come from the contribution of solar power, wind power, and natural gas. Nuclear and hydro energy are projected to grow as well but not significantly. Since oil, coal, and natural gas will remain the main contributors to the energy supply, they will dominate the energy landscape, feeding around 75% of the world's demand. Meanwhile, greenhouse gas emissions are foreseen to fall by 2035 because of higher engine efficiency and a shift toward natural gas and renewable energy. While this is good news, it is not enough to fight climate change.

---

[62] (McKinsey & Company, 2016)
[63] (Mona Hammami, 2016)
[64] (International Energy Agency, 2017)

Technology will need to play a role in the upcoming decades. How technology improves the efficiency of each energy source will shape the energy mix of the future. The challenge for economic and political leaders will be to collaborate at all levels of society to lead action toward a correct balance of energy sources. Clearly, the world is still very far from being sustainable energy-wise.[65] [66]

Prosumerism, the hope of society?

The current system is based on a one-way direction. Let's take the example of electricity. It is produced in power plants (mainly from fossil fuels, as we saw in the last section), sent through a transmission grid to the cities at high voltage, distributed to the neighborhoods at medium/low voltages, and consumed in each building and house. Electricity makes a long trip to arrive at the lamp that is illuminating your room. This system implies energy loss,[67] but it is straightforward to manage and protect. Usually, the production, transmission, and distribution companies are few (if not singular[68]) and they are the ones that control the grid. Each player is stable and has a definite role.

The new industrial revolution has the potential to change this system. People will shift to more collaborative ways of locally producing and distributing energy in ecologically friendly ways. The trends are moving toward the idea of *prosumerism*. A *prosumer* is a producer who is also a consumer. The concept is simple. Each house, building, and business will generate its own green electricity from solar panels on the roofs, geothermal power from the ground, waste-to-energy systems, or even a small wind turbine hidden in the backyard. In Germany alone, there are currently more than a million buildings generating their own

---

[65] (McKinsey & Company, 2016)

[66] (Salvador Lopez, 2016)

[67] When electricity moves through a wire, heat energy is produced and lost, implying that the final energy available is less than the starting energy.

[68] In the majority of developed countries, transmission is supervised by a single government-owned company that has a monopoly on the high voltage grid, and distribution is controlled by several companies that compete in a liberalized market (usually creating oligopolies).

green energy. They receive a royalty if they return the extra energy to the grid. This fee depends on supply and demand. At dinner time, the electricity price reaches its peak because there is no solar energy available to generate electricity (high demand + scarcity of supply = increase in price). The prosumers who have the most energy available are in demand. Therefore, small storage players will appear with new technologies (mainly lithium batteries) to stabilize the market and take advantage of these peaks in demand. They will capture the power during the day and release it when the grid needs it (and, of course, make a profit out of it). A new peer-to-peer communication will be needed in the whole electric system to manage this market. Thanks to the internet and the explosion of sensors, setting up a small grid with your neighbors will become easier. This new system will not have a few players; it will have many. It will not be one-way, but multi-way. It will not be top-down; it will be collaborative and peer-to-peer. At some point, the whole high-voltage transmission network will simply become outdated because every single consumer will also be a small producer attached to a local grid.[69] [70]

The concept is beautiful. However, even if we ignore all the technical implications and risks from a prosumerist society,[71] the reality is that this new way of producing and consuming energy will take decades (if not centuries) to be fully implemented. We sometimes tend to believe that when innovations are discovered, they invade the world within a matter of years. Electrification started in the mid-1880s, but even today there are 1.1 billion people without basic access to electricity. Innovations that change the world take time. However, prosumerism is the path to follow if we want to live in a more sustainable world.

Tracking energy consumption

---

[69] (Rifkin, 2015)

[70] (Big Thinking, 2016)

[71] None of the electric grids in the world are ready. Some investments need to be done, especially from the points of view of protection and stability.

As we already mentioned, sensors are one of the key features of this revolution. The energy sector is no exception. The current trends show that tracking where, when, and how energy is consumed in your home or business enables costs savings that can reduce your energy bill (and help combat climate change). A rising number of eco-friendly solutions have appeared on the market.

Smart meters are digital utility meters installed by the power companies or by individuals. They send wireless signals to show the amount and type of electricity your house is using. In the United States, some power companies have already started to install smart meters, making it easy for them to check your energy usage remotely and in real time to make energy-saving suggestions. In Europe, the power companies are still on hold for these smart meters, but you can privately install systems to monitor your electricity usage behavior. As you might expect, smart energy apps are at the end of all this information and make it easy for you to understand how you consume electricity and what you can do to reduce your energy consumption.[72]

Another example. Tracking systems using SCADA[73] technology are used by the TSOs[74] and DSOs[75] to optimize energy production systems. For example, the system might start a gas turbine from a power plant when the sun is not bright enough to produce solar energy and the grid needs more energy. This automatic monitoring will be able to control the whole electric grid more efficiently, reducing costs and emissions.[76]

The impact of these monitoring systems will not stop at electricity. Buildings will be equipped with SCADA systems that will make

---

[72] (Digital Trends, 2016)

[73] SCADA stands for Supervisory Control And Data Acquisition, which is a control system used for supervisory management of high-level processes.

[74] A TSO is a Transmission System Operator, a company in charge of transmitting electrical power from generation plants over the electrical grid to regional or local distribution operators.

[75] A DSO is a Distribution System Operator, a company in charge of carrying electricity from the transmission system to individual consumers.

[76] (Siemens, 2013)

them *alive*. Technology allows for 24/7, real-time monitoring of heating, power system, lighting, ventilation, etc. Therefore, the building will be able to adapt its energy use depending on the number of people inside the building and the weather conditions of the day. The building will *react* to each new situation. These technologies, combined with the use of better construction materials, will reduce the overall energy bill.[77]

Last but not least, transportation will also be monitored to reduce energy consumption. Fuel consumption monitoring tools are software systems that track data over time and compare patterns. Some companies have developed in-house software and offer plans for their clients to be more energy conscious. Computational requirements are modest and cheap, as the analysis is not sophisticated. Therefore, these investments are quickly offset by savings. The best thing is that such systems will provide real-time feedback with recommendations that will reduce costs and emissions.[78]

Final thoughts

The energy sector is facing an industrial revolution too. We will definitely produce more efficiently in the next few decades. But remember, the trends show that these changes will still take time. The speed of disruption will differ for every country. Our world is still hugely dependent on fossil fuels; they still represent more than 70% of our overall energy mix. There is hope, however. With the new efficiencies that this industrial revolution will bring about and with the shift in our mindset regarding where our energy comes from, we are prepared to take big steps to fight climate change. We will use sensors to monitor where our energy comes from, how it is distributed, how we use it, and what emissions we generate. Further, we will be able to track our carbon footprint, be more aware of our impact on the environment, and live more sustainably.

Box 1: Solarly becoming the lights in the night

---

[77] (Technavio, 2016)
[78] (World Bank, 2016)

Solarly is a start-up bringing hope to the future of energy in developing countries. In sub-Saharan Africa, 2 out of 3 people do not have access to electricity. The national grid is usually unstable and limited to specific regions. That's a problem that directly impacts the development of the country in every layer of society and every sector of the economy: from education to security, from health to business. Solarly's solution to this problem is to provide easy access to electricity to rural households with an off-grid, autonomous, and affordable solar energy system. Solarly sells a solar station plan that includes photovoltaic panels, a storage battery, USB connectors, 12V DC connectors, and a remote management system. Each solar station can be connected to others to create a grid that is independent of the national grid. Since panels are charged during the daytime, citizens of the community can continue their activities even after sunset. The users pay via SMS technology, and the income generated provides a business model to the person who maintains the grid. This company represents a local and simple solution for a global problem that is affecting millions of people. That's the inspiring part of technology: making affordable solutions that bring life-changing value to communities.[79]

Box 2: ProsEU mainstreaming active participation of citizens

With the objective of fulfilling the Paris Agreement, the European Union is investing in a project called ProsEU (PROSumers for the Energy Union) under the Horizon 2020 program. The idea is to provide citizens with the tools to take ownership of the transition toward greener energy by enabling citizens to become real prosumers. The aim is to close the gap between investors, producers, and consumers. ProsEU is a three-year research project designed to understand the incentive structures needed to enable new business models to attract investors and citizens. Don't forget: market regulations, infrastructural integration, technology scenarios, and governmental policies play a central role in the energy sector and are sometimes barriers to the development of sustainable energy. It is an example of a solution

---

[79] (Solarly, 2016)

that could be feasible in the future EU market. Time will tell which business models are chosen to manage the way we fuel our need for energy. Nevertheless, it is a good beginning.[80]

Box 3: When wind goes offshore

In renewable energy, one big hope is offshore wind energy. The European Union in particular has experienced substantial offshore wind power expansion in recent years, reaching 11,027 MW of production. Including sites under construction, there are 84 offshore wind power plants in 11 European countries. And the trend is not just limited to Europe; the rest of the world is catching up too. China had a total of 226 offshore wind turbines in 2014 and is speeding up its development in the area, while the United States came onboard in 2016 with its first offshore wind farm off Rhode Island. Such global growth has helped drive investment in wind power, which has led to a decrease in prices by nearly 60% since 2000. The decreasing price trend is also seen in other renewable technologies and represents a pattern of new technologies appearing. Buying clean energy is becoming cheaper and cheaper, and, as surprising as it may sound, China is leading the way, despite being the world's worst polluter as well.[81]

---

[80] (Drift for transition, 2018)
[81] (Wind Europe Association, 2016) and (Siemens, 2017)

iv. Financial services
To begin with

Banking and financial services, in general, are going through times of change. With notable bankruptcies like that of the Lehman Brothers (which accounted for assets worth $600 billion, the same as the GDP of Argentina in 2017[82]) the financial crisis of 2008 collapsed the banking industry. Traditionally, the banking sector is based on trust, so the damage to its reputation was a hard hit. At the same time, some new *fintech* companies were popping up, proposing new business models that could potentially disrupt the world's understanding of banking and money. A fintech is a start-up that aims to provide financial services by making use of software and modern technology. For example, PayPal changed the way people could make payments online with an intuitive and secure system. Even the entire central banking system was challenged with the introduction of cryptocurrencies like Bitcoin.

Everyone is talking about how fintech might shape the world. Fintech, just like banking, is everywhere. Every business involves transactions. Therefore, the potential disruption of fintech technologies is global and spans across sectors. The hope is that these technologies will bring financing options to people without banks, reinvent digital models, rebuild the whole value chain, and even bring a new paradigm to our old tax collection systems. Table 04 describes the leading trends in the market.

---

[82] (International Monetary Fund, 2017)

WEALTH MANAGEMENT	Robo-advisory, social investing, crowdfunding …
INSURANCE	Telematics, social integration, IoT and connected devices, prevention …
CAPITAL MARKETS AND INVESTMENT BANKING	Next-generation trade finance, collateral management, trade analytics …
SMALL AND MEDIUM ENTERPRISES	Peer-to-peer lending, digital cash, one-stop shop for businesses …
PAYMENTS	Mobile payments, international remittances, mobile point-of-sale devices cryptocurrencies, …
RETAIL	Personal financial management, peer-to-peer lending, investment, aggregator comparison engine …

TABLE 04: EMERGING AREAS THAT ARE BECOMING NEW NORMS IN BANKING[83]

You may think that any of these examples can fuel a book by itself. And you are right.

However, I would like to focus on the three main innovations (in my opinion) that will reshape the financial system: blockchain, AI advisory, and crowdfunding.

Blockchain will change the industry

Blockchain technology is an open-source distributed database utilizing high-class cryptography that can alter the structure of the global economy through collaborative monitoring of every single interaction and transaction. Sounds complex, right? Well, it is much more complicated than it sounds. Keeping it simple, you have to remember that:

- Transactions can be done without the involved parties knowing each other's identity and without the mediation of central authorities and classic banking intermediaries (such as VISA or your retail bank)

---

[83] (McKinsey & Company, 2016)

- The use of cryptography makes blockchain very secure.
- Blockchain allows anyone to rewrite the code. However, any change has to be approved by the whole system.
- Bitcoin is a cryptocurrency based on blockchain, but not the only one. There are more than 2,200 other cryptocurrencies at the time of writing this book.[84] The most popular after Bitcoin are Ethereum, Ripple, and Bitcoin Cash.
- Investing in cryptocurrencies is a gamble. The market is extremely volatile.

Blockchain technology is bound to revolutionize the financial industry and influence how transfers, savings, trades, loans, and all other processes involving money work. It promises a more progressive world where people can participate in any aspect of their society. However, the present challenges are the vast amount of energy this technology consumes, the adversity that will be caused by smart agents replacing regular employees with their jobs, and governance issues. Since blockchain technology is still in an early stage, there are no systematized actions as yet to address these challenges. As usual, legislation is taking a long time to develop. Blockchain can dramatically restructure business models, markets, risks, and savings of cost and capital.[85] Some applications worth knowing are:

Smart contracts: Almost any intangible document or asset can be expressed in code, which can be programmed into or referenced by a distributed ledger. Blockchain technology has the potential to digitalize contracts because you can record something on it and, once registered, it is an irrefutable digital proof that this thing happened at this time on this date between these counterparties. The smart-contract can be anything, from a marriage to a divorce, from the purchase of a diamond to an insurance claim for the loss of that diamond, from a house sale to a land reclamation. Access to real-time trade details would enable digitized smart contracts to be verified instantaneously,

---

[84] (Investing.com, 2018)
[85] (McKinsey & Company, 2016)

assuming that pre-defined conditions are met. It would allow a letter of credit to be issued more efficiently than in today's trade finance process.[86] And more importantly, checking the authenticity of the letter of credit would become much more efficient than the current paper-intensive process. There are already new projects such as Ethereum and Blockstream that allow for comfortable, secure smart-contracts and digital asset management.[87]

Leaner banking structure: Distributed ledger technology could reduce banks' infrastructure costs attributable to cross-border payments, securities trading, and regulatory compliance by $15-20 billion per year by 2022. The current financial system involves a large number of manual checks that must be processed to verify the legitimacy of a transaction. Most of these checks require the presence of a physical person, a cost that doesn't bring real value to any bank.[88]

Cryptocurrencies and payment systems: Bitcoin is the most popular cryptocurrency based on blockchain, but there are many others. You may have heard of Ethereum, Ripple, Litecoin, or Monero. These new digital, decentralized currencies can change the way we make transactions. Currently, when you use your VISA card in a restaurant, a transaction fee of around 3% is levied for the business and maintenance of your credit card service. With Bitcoin, you only need a Bitcoin wallet (with bitcoins in it, of course). You can send bitcoins over the internet to anyone. You do not need a bank to make a transaction. Moreover, Bitcoin transactions are almost free (with a few cents going to the network as miner fees), and you don't have to pay monthly machine rental fees or deposits. The main downside of Bitcoin is the limitation of the number of transactions that can be performed per hour. However, there are new, lighter currencies such as Ripple or NEM that want to become the mainstream currency. Besides, second-layer scaling innovations such as the

---

[86] (Skinner, 2016)
[87] (Prisco, 2015)
[88] (Santander InnoVentures, Oliver Wyman and Anthemis Group., 2015)

lightning network aim at making Bitcoin, along with other cryptocurrencies, competitive with mainstream payment processors.[89]

Mortgage industry: Blockchain technology will likely alter the process through which consumers buy a home, as well as the way financial institutions handle mortgages. Explicitly, the technology could remove cost and friction from the process, create transaction records that are infallible and incorruptible, and facilitate near-instantaneous settlement. It could also dramatically change the way mortgages are serviced and sold on the secondary market. First, blockchain might help establish more accurate record keeping. At fulfillment, smart contracts would speed up settlement flows. In the mortgage-servicing process, blockchain could track the movement of payments, and in the secondary markets, it might provide transparency about the ownership of underlying assets. Financial institutions will have to adapt and change their core infrastructure if they want to compete in the fast-changing market.

The real potential of this technology is yet to be discovered. The doors that blockchain is opening are infinite and exciting.[90] [91] [92]

AI will become the name of your banker

Artificial Intelligence (AI) used to be just an idea from science fiction books that described the next step in human development. However, it is now becoming a reality and has already been applied in various sectors and industries. AI's potential is huge, and it may possibly destroy a lot of jobs. AI is built with a combination of algorithms that is able to learn from humans. We believe that machines are extremely powerful, and they are—but only when you use them for specific tasks. For example, a machine is great at doing repetitive calculations quickly and accurately. In this area, they outperformed humans

---

[89] (Verhelst, 2017)
[90] (PwC, 2016)
[91] (Deloitte, 2015)
[92] (Verhelst, 2017)

long ago. But they were bad at "thinking" and "understanding." They were simply not able to learn from humans (or we were not able to teach them)—until now.

New algorithms have been developed to make computer "learning" possible. Computers are now able to see patterns and draw conclusions by themselves. AI is a technology that can inspire thousands of books, but I decided to focus on the added value that it can bring to the financial world.

For many years, banks have been hiring data science specialists, statisticians, and mathematicians to conduct risk analyses, financial assessments, and investments or portfolio organization. Now AI can help by providing accurate insights for decision-making when it comes to starting a business, applying for loans, or making investments. AI is used in robo-advisers/bankers that can assess risk and reply to customer questions almost instantly. Some applications are already taking advantage of this technology and show a high potential for the years to come. These are:

Enhanced Customer Personalization: Online wealth management services that provide automated, algorithm-based portfolio management advice are the key to genuinely personalized financial help (unlike the boring impersonal emails that our bank sends us every month about the new products that we do not want or need). Customers will hopefully come back to the core. AI is highly and cheaply scalable, whereas humans are expensive and like to change jobs from time to time.

Productivity gains: Forget about bank offices and retail bankers. The new generation of customers doesn't use them. From customer communications to necessary back-office processing, AI can take the classic boring and repetitive processes and make them both efficient and effective. What once was a tedious process of signing up for a bank as a new customer can now be done in a few minutes with your mobile phone in a highly personalized interaction. This level of personalization and productivity is impossible to achieve without the benefits of machine learning and AI.

Fraud detection: Fraud detection is especially important for insurance companies. Fraud is a classic risk to take into account. I have good news: AI can help. By monitoring and reviewing account activity patterns, deviations from routine behavior can be quickly flagged for further review. Over the last decade, AI has not only significantly improved the fraud monitoring process, it has also responded in real time to potential fraud. Artificial intelligence can now detect if suspicious transactions are being made by running algorithms that can detect fraudulent patterns of user activity. Moreover, these algorithms can quickly and automatically suspend the accounts of suspicious users, helping banks to save money and reduce crime.

Soon, all financial service firms will leverage the power of AI to deliver better customer experiences, lower the costs of the industry, and reduce its risks. Companies are still figuring out how to restructure themselves to leverage AI capabilities. Just as building a website doesn't make you an internet company, sprinkling on a little bit of machine learning doesn't make you an AI company. Even if banks keep some of their branches alive to serve the more traditional (and wealthy) members of society, the industry will become more and more digital.[93] [94] [95]

Crowdfunding and access to capital

Much to the surprise of bankers, small businesses and start-ups now have somewhere else to turn if they are rejected by the bank. It is common knowledge that banks are cautious when it comes to lending money to start-ups. And that's understandable. The banking business is based on assessing risk and lending money. Start-ups and small businesses tend to come with huge risks and low potential returns in the short term.

---

[93] (McKinsey & Company, 2108)
[94] (BBVA, 2016)
[95] (Marous, 2017)

Luckily for start-ups, peer-to-peer funding platforms are emerging. Kickstarter ($2 billion raised since 2009), Indiegogo ($1 billion raised since 2007), and CircleUp ($300 million raised since 2011) are the most famous platforms so far. Altogether, peer-to-peer funding raised $34 billion in 2015.[96] Moreover, new ways of raising money such as ICOs[97] are becoming more popular in the market with estimates stating that ICOs raised funds worth $6.3 billion in the first quarter of 2018.[98]

But crowdfunding is more than simply investing in cool ideas that founders present to you on a digital platform. Funding Circle, for example, aims to provide more proof of a business's credibility by allowing its users to speak directly with the business owner on a one-on-one basis. This strategy is convenient for new borrowers since it's less tedious than the bank's process, and they can have the money in their hands in a shorter time and at a lower interest rate. More and more companies are venturing into offering business loans, including Kabbage, Square, and even PayPal.

Some small companies that do not have access to affordable credit are now able to monetize one of their biggest assets: the money customers owe them. Normally, small businesses have to wait for 30-90 days after they have dispatched merchandise to get paid. This cash-flow problem can limit their expansion. Financial players have traditionally offered "factoring," a process that allows financial players to buy outstanding invoices at a discounted rate—sometimes a hugely discounted rate. The paperwork involved in factoring is daunting. However, by moving invoices onto electronic platforms, fintech start-ups hope to make the process frictionless. If e-invoicing became the new norm (and it likely will), online factoring would be easily possible. If a local

---

[96] (Investopedia, 2017)

[97] ICO stands for Initial Currency Offering. It is a fundraising mechanism in which new projects sell their underlying crypto tokens in exchange for bitcoin and ether. It's somewhat similar to an Initial Public Offering (IPO) in which investors purchase shares of a company. With few regulations and remarkable ease of use, this ICO climate has come under scrutiny from many in the community as well as various regulatory bodies around the world (Nasdaq, 2017)

[98] (Coindesk, 2018)

business sells a shipment of ball-bearings to BMW and the carmaker agrees to pay the invoice electronically within six weeks, the invoice is nearly as valuable as a BMW bond! It might be worth 98 cents per dollar and not 60 cents per dollar. The verified invoice can then be auctioned on a factoring platform and can give start-ups liquidity to expand further. By turning the invoice into a fungible security in a flexible and quick manner, the local business in effect piggybacks on BMW's credit rating, which is likely to be much better than its own. It's a win-win situation and another way of overtaking banks.[99]

Final thoughts

In summary, the financial industry will be massively disrupted in this industrial revolution. Currency as we know it will change, as will our understanding of its value. The need for banks in some of its core business models will be challenged, and big data combined with artificial intelligence will be able to provide excellent customer service to satisfy our financial needs. What is cool is that all these innovations will not only change how we trade and insure our risks, but they will also completely reverse the status quo between the financial industry and the citizens of a country. Central banks issuing currency and large banks trading it will no longer hold a monopoly on the money supply in our economies. Your private financial data will no longer be gathered in the central server from your retail bank. You will have access to new sources of financing to build your dreams. We will realize that money's value is, by definition, linked to the value that a community gives it. The way we trade this value doesn't matter. What is important is that the system is reliable enough and has the desirable ethical values. This industrial revolution will offer us the choice we deserve to reach our financial freedom.

Box 1: Ewala and the access to financial services

Ewala (ewala.co) is a promising Belgian start-up that offers instant international mobile transfers with low fees using your cell phone

---

[99] (The Economist, 2015)

number. The World Bank estimates that officially recorded remittances[100] to developing countries amounted to $429 billion in 2016. The main players in the industry are Western Union, MoneyGram, and RIA, and they levy substantial fees on their customers for transferring money from one country to another. Ewala is another example of all the new fintech players that are entering the financial market by offering an innovative solution to the new (or old) problems that society is facing. The whole financial industry will have to evolve and adapt if it doesn't want to disappear in the upcoming years.[101] [102]

Box 2: Kensho is becoming the Siri of the financial industry.

Kensho is a market data analytics company that can find answers to more than 65 million questions by scanning over 90,000 customizable actions. Questions, asked in plain English, are typed into a Google-style text box. Kensho's creative approach to data analytics allows every user to request information about almost every aspect of the financial industry. Traditionally, these financial questions would require several analysts and several days to be answered, but with Kensho, these questions can be solved in just a matter of minutes. Kensho was founded by MIT and Harvard in 2013 and has a pool of expert engineers, UX designers, and data scientists who are focused on creating software that studies unbelievable amounts of data and finds patterns and solutions in it. In November 2014, the company received $15 million in new investment. Using big data and analytic technologies as tools, Kensho has become the most significant financial analyzing company in the world and has the ambition of becoming the next billion-dollar company. Because of its capability, it is now used by giant financial firms such as JP Morgan and Bank of America.

---

[100] A remittance is a transfer of money by a foreign worker to an individual in their home country. Money sent home by migrants competes with international aid as one of the largest financial inflows to developing countries. (Wikipedia, 2018)
[101] (World Bank, 2017)
[102] (Ewala.co, 2018)

Who said that virtual assistants were only for business-to-customer clients?[103]

Box 3: MaidSafe for secure P2P information storage

Do you remember when you used to download music and videos with programs such as LimeWire or uTorrent? You were using a peer-to-peer technology. You got great speeds for your downloads but, in exchange, somebody else was using your computer as a server. Rather than using centralized data centers, we were using a network of computers. MaidSafe uses the same principle. The SAFE Network uses advanced peer-to-peer technology that joins together the spare computing capacity of all SAFE users to build a global network. In other words, SAFE is a crowdsourced internet storage service. Everyone's data and applications reside on this network of computers. Imagine having a service like Dropbox where no one could spy on you or take down your data (which happened in 2012, when a cyberattack stole over 68 million users' email addresses and passwords and leaked them onto the internet[104]). When you donate your unused storing resources to SAFE, you will be paid in a network token, called safecoin. Over time, your computer will start to fill up with network data, and as a consequence, your virtual wallet will automatically begin receiving safecoins. You can use this currency to pay for other services on the network or convert it to another currency via a SAFE currency exchange.[105]

---

[103] (Forbes, 2015)
[104] (The Guardian, 2016)
[105] (MaidSafe, 2017)

Hollywood loves to act out the end of the world due to evil robots that control defenseless humans, but it is a fact that societies are quickly moving toward a digital world. From agriculture to financial services, we have seen that industrial disruption is unstoppable and will change society. It won't be as fast as we might expect due to the cultural challenges attached to change and innovation. However, the people who know how to take advantage of automation and digitalization will stand to gain in the foreseeable future. Technology will be The Enabler of the twenty-first century.

B. Will you become Superman?

When we think about the next industrial revolution, we tend to only take into consideration the changes in the material world: more stuff and better connectivity. However, this revolution goes much further. One of the most exciting changes brought about by technology will be in the biological world. This revolution will provide us with tools that will change our understanding of what a human being is and can do. Will we be able to play God?

i. Mastering your genes
To begin with

Let's begin with a short baseline to put everyone on the same page. Your body is formed by cells, approximately 37 trillion of them! Crazy, right? In general, human cells—including those found in your eyes, heart, bowels, and other organs—have a nucleus.[106] Inside this nucleus, you will find the majority of the cell's genetic material. This material is formed of DNA molecules[107] along with a variety of proteins that form chromosomes. In humans, each cell usually contains 23 pairs of chromosomes, amounting to a total of 46. Each chromosome contains many genes (humans have between 20,000 and 25,000 different genes). Genes act as instruction manuals to make new molecules called proteins. Those genes, my dear friend, define all your traits,[108] from the color of your eyes to your likelihood of suffering from Alzheimer's when you grow up. Imagine the possibilities if we could control and monitor those genes... but wait, can we?

---

[106] In nature, there are two types of cells: prokaryotic (without the nucleus) and eukaryotic (with nucleus). Even though we have gobs of prokaryotic cells living inside us, humans are still categorically eukaryotic organisms.

[107] DNA means DeoxyriboNucleic Acid (cool name for an acid). It is a molecule that carries the genetic instructions used in the growth, development, functioning, and reproduction of all known living organisms.

[108] A trait is an inherited characteristic which is found in the phenotype (the physical appearance) or genotype (the genetic material) of an organism. Traits include eye color, hair color, the shape of body parts, genetic diseases, etc.

Building better humans

Despite criticism, genetic engineering has been going on for centuries. Early on, we learned to identify the healthiest and strongest plants and animals. Through selective breeding (a rudimentary form of genetic manipulation), we became capable of reproducing desirable traits and amplifying them in the next generation. If we compared a cow from 100 years ago to a cow living today, we would definitely be surprised by how different they are. Genetics, genetics, genetics…

Although we are still learning how to efficiently and safely repair and modify bodies, including our genomes, we have been reading and writing genetic codes of other organisms in the laboratory for decades. Sounds crazy, right? Biotechnology allows humans to make genetic improvements. Governments and companies have been continuously and generously funding research focused on the manipulation of genes; there are billions of dollars in the biotechnology industry. Currently, people are willing to spend large amounts of money on products like wrinkle prevention creams, anti-aging pills, workout enhancing drugs, breast implants, butt implants, cheek corrections… there is a clear business model here, and genetic modification could achieve the same (or even better) results than the products currently on the market! But looks are not the only thing that can be improved. There is something that affects the rich and poor alike: aging. Money can buy you a paradise, but it can't buy you the youth to enjoy it. Whoever finds the long-sought Holy Grail of immortality will change the course of humanity forever. There is strong reason to believe that humankind can produce cells that are completely resistant to all known viruses.[109] It should also be possible to engineer other traits, including reducing the likelihood of having cancer or other diseases. We're not there yet; creating virus-resistant human cells won't be easy, and it could take a decade or longer to do it. However, I'm sure you'll agree with me that the outcome will be worth the effort. By producing safe, virus-proof human cell lines, scientists would be able to

---

[109] (Time, 2018)

produce drugs, vaccines, and antibodies without the risk of viral contamination. Moreover, the chance of a person's immune system rejecting new drugs will be lowered. The main stopping point is that synthesizing DNA is expensive. Currently, costs are about $1 for 10 DNA base pairs (and humans have 3 billion DNA base pairs!). However, the price is going down and will hopefully reach an amount that will allow more projects to emerge. Since 1985, costs for reading and writing DNA have plummeted, while the hourly throughput of instrumentation has improved exponentially, doubling every 18 months or so. The Human Genome Project-Write, GP-Write, and Project Recode are just some ambitious examples of what genetic engineering is able to do. Believe it or not, we are already building better humans, and the trend is just moving forward.

But a longer lifespan is not the only boon biotechnology has to offer. It can also help us create better kids. Let me narrate a story that might some day (depending on your age) sound familiar to you. Once upon a time, there was a couple who wanted to have a daughter, but both partners were affected by a genetic disease. After nine months and lots of genetic prescreening, a healthy and beautiful baby girl named Jennifer was born. This success came through in vitro fertilization. Doctors not only chose the sex of the child, they also used their expertise to diagnose and prevent deadly genetic diseases at the embryonic level. Cool? Scary? It's up to you to decide. However, this story is reality. British scientists have developed a "genetic MoT" test. This technique offers a universal method of screening embryos for diseases by using karyo-mapping, which is more efficient than previous processes. This method costs $3,000 and could be a game changer for the industry. The test is performed on a two-day-old IVF embryo and allows doctors to screen for gene combinations that create high risk of diabetes, heart disease, or cancer. Impressed? This is what we are able to do now. In the future, we may also be able to "cure" genetic diseases in embryos by replacing faulty sections of DNA with healthy DNA, a process called germline therapy. This has been performed in animal embryos but is currently illegal for humans. Taken to their extreme, the developing technologies of genetic alteration might open up a whole new set of possibilities that could allow us to simply design babies as we want. [110] [111] [112] [113]

CRISPR-Cas9 and biohacking

In the old days, when scientists wanted to alter a cell's DNA, they had to deal with a complicated and costly process that had low probabilities of success. Today, they just use CRISPR.

CRISPR joined the medical landscape of DNA modification in 2012, when scientists Jennifer Doudna and Emmanuelle Charpentier published their work in *Science* magazine and disrupted the medical world forever. What is CRISPR? I'll give you the short answer and the long answer.

The short answer is that CRISPR is a new class of molecular tools that scientists can use to precisely target and cut DNA. This technique allows scientists to edit precise points on the DNA strand using a bacterial enzyme. It is fast, easy, and cheap.

The long answer is that CRISPR stands for Clustered Regularly Interspaced Palindromic Repeats. CRISPR units consist of a DNA-snipping protein and a genetic GPS guide. CRISPR units have a code that complements a specific part of the DNA, so that the unit can only attach and cut the DNA at a specific, complementary location. The snipping proteins are naturally occurring and usually prevent DNA from getting tangled. CRISPR systems naturally evolved across the bacterial kingdom as a way to recognize and disable invading viruses. But researchers recently discovered they could repurpose that primordial immune system to precisely alter genomes. In other words, you can play God with your DNA and with the DNA of humans yet to be born. We will be able to hack our own bodies. If you are not yet impressed by this innovation, let me describe some real

[110] (Carlson, 2017)
[111] (Gizmodo, 2018)
[112] (The Telegraph, 2009)
[113] (CNN, 2008)

applications that are currently being developed by biotech companies (this list is just a small portion of all the research going on):

Mosquitoes carrying malaria could be genetically edited so they no longer pass on the disease to future generations, saving millions of lives in the future.

Humans could eradicate certain genetic diseases, like cystic fibrosis. By genetically engineering the sperm cells and egg cells of the parents, the transmission of certain genetic diseases to their kids can be completely prevented.

Liver cells could be edited so that they lower the bad cholesterol in the blood, reducing the risk of heart disease, which is currently killing millions of people every year.

Plant-based building materials that are lighter than aluminum and stronger than steel could be developed and used.

Antibiotics that only target bad bugs without wiping out the entire microbiome could be synthesized.

Biological control in our ecosystem could be realized by wiping out an entire species in the name of conservation. For example, in Australia a non-native species of frog that was carried into the country wreaked devastating effects on the local ecosystem. It could be possible to sterilize all those animals and bring back the native biodiversity.

So far, though, very few CRISPR-enhanced products have made their way into the hands of consumers. Although CRISPR isn't going to end disease or hunger or climate change any time soon, it is already beginning to reshape the physical world around us in much less radical ways, one base pair at a time. The regulation and the ethical debate are, again, moving slowly. The ethical debate is inevitable if we want to take advantage of all the good that this innovation can bring to humanity.[114] [115]

---

[114] (Wired, 2018)
[115] (Future Today Institute, 2017)

Time to clone humans and build a super army?

Creating a genetically identical duplicate of an organism is now possible through cloning. Cloning describes a number of processes used to create genetic copies of a biological entity. There are two types of human cloning that are often confused with each other: therapeutic cloning and reproductive cloning. Therapeutic cloning involves cloning cells from a human for use in medicine and transplants. Reproductive cloning involves making an entire cloned human, instead of just specific cells or tissues.

So how exactly can we clone a biological entity? The technique is old. It is based on somatic cell nuclear transfer. It involves removing a mature somatic cell (any type of cell, except sperm or egg) from your body and transferring the desired DNA into an egg that has had its own DNA removed. This creates an embryo, which is implanted into a surrogate mother through in vitro fertilization. *Et voilà*! Nine months later, a new person will be born with the exact same DNA as you. Easy, right?

Well, the process is far from perfect. The steps described only make 1 viable embryo in every 200 attempts. Even if this embryo makes it though the process and the baby is born, the results can be very unpredictable, making the whole process hard to justify. Let's take, for example, the case of Dolly the Sheep, the first mammal cloned from an adult somatic cell. As a clone, her lifespan was only about half that of a normal sheep. Scientists haven't yet found a technique to create clones that are as healthy as their natural counterparts, but with new technological innovations, they will. It is simply a matter of time.

Cloning, despite the ethical controversies surrounding it, can have many applications in the development of our societies.

The first is the possibility of cloning healthy cells. As humans, we create cells continually until we die. With each cellular division, the DNA that is transmitted to the new cells undergoes more mutations and accumulates more damage. Through cloning, those damaged cells (which are more prone to disease) could be eliminated and replaced with healthy cloned ones.

Moreover, working on cloning techniques brings us a better understanding of who we are and how we grow by providing important insights into human embryonic development.

But probably the most promising potential of cloning research is providing solutions to many developmental diseases and birth defects, with stem cell therapy or regenerative medicine. Imagine having a copy of your DNA ready to regrow any organ or tissue you may need with very low risk of rejection from your own body. For example, a person with a liver disease could potentially "grow" a new liver using his or her own genetic material. How much would you be willing to pay? We can even go to the next level! Imagine that we could commoditize organs and have a market for them. Currently, an estimated 21 Americans die every day waiting for transplants.[116] Imagine having a large stock of livers available in each hospital, just waiting to be transplanted. Commoditizing life might sound crazy, but we already do it. Like it or not, illegal organ donation systems exist and always will. From China to Germany to the United States, scandals have gone public showing the world that money can buy organs. The development of cloning could potentially change how we understand organ donation and the value of a human body. The framework and ethical boundaries that we set for our societies will be the key to defining the expansion of this technology. The future of cloning is in our hands.[117] [118] [119] [120]

Final thoughts

In a nutshell, technology will allow humans to live longer with a better quality of life. Biological advances will not only change the way we live and procreate, they will also change the way we

---

[116] (Wen, 2014)
[117] (Connolly, 2013)
[118] (Walla, 2006)
[119] (Samuels, 2016)
[120] (Future Human Evolution, 2017)

understand life and death. It is true that we are still in the early stages. However, there are a lot of projects that are giving inspiring and positive results. We are managing to play God.

The main barriers are on the ethical side. The legal frameworks that governments build around biotechnology will define its success or failure. As usual, with this new era, we will witness a moment in history where we will face a trade-off: we will have to decide if we want to move forward with biotechnology, despite its risks, or if we should stop its development and preserve our current definition of life. Does the end (saving lives) always justify the means (experimenting with human beings and uncertain results)?

Box 1: Gentle and your DNA

With the innovations available in the market, genome companies can provide you with access to your genome sequence for less than $2,000. Everyone can access their genes for an affordable price. Gentle Labs was born with the mission of reducing the unnecessary suffering caused by genetic diseases that can be detected at an early stage of embryonic development—or even in healthy parents. It is determined to find more correlations between genes and complex diseases. This company has developed the first consumer genetic test based on the sequencing of all your genes. Gentle's analysis is much more accurate and complete than its competitors' analyses, enabling users to make scientifically sound conclusions from the tests. Moreover, the scientific community is discovering new correlations every year, ranging from disease patterns to drug responses. Gentle has the aspiration to keep its clients informed about scientific progress, enabling them to apply new knowledge to their DNA directly.[121]

Box 2: China's $9 billion effort

The United States has always been the world leader when it comes to genomic research and tailored healthcare with DNA.

---

[121] (Gentle Labs, 2017)

Expensive researchers hope to find solutions to genetic problems through altering, modifying, and/or entirely engineering DNA. The potential profits of discoveries in this domain are huge and could bring a significant competitive advantage to any country that makes advances.

China is sinking billions of dollars into research and development, promising projects both in China and abroad. One of the main investments is the Chinese company Wuxi NextCODE, which is based in the United States. China believes that this technology could prove transformational and give a clear competitive advantage in their path toward world dominance. The genomics revolution could be China's chance to showcase its technical prowess. Over the last few decades, China has transformed itself into an economic superpower through massive industrialization. However, amidst slowing growth, their industrialization is now facing limits such as a shift of the manufacturing workforce to cheaper nations or the consequences of overpopulation and contamination. Chinese companies have now become top players in the biotechnology field. Just like the USSR-US competition to colonize outer space during the Cold War, the new battlefield seems to be the world of genetics. Time will tell who will win this race.[122]

Box 3: The BabySeq project

In developed countries, newborn screening is required for babies. Such testing can predict a few genetic diseases, but it's far from providing a full picture. That's why the Baby Sequence project was born.

The BabySeq project is designed to examine how best to use genomics in clinical pediatric medicine by creating and safely testing methods for integrating genetic sequencing into newborn care. Unlike the existing newborn screening, which can only predict up to 30 diseases, BabySeq will look into 1,700 protein-coding genes that are strongly tied to illnesses that begin during

---

[122] (The Washington Post, 2016)

early childhood. The objective of this project is to search for mutations connected to specific diseases and genes associated with incurable conditions that occur during childhood. Thanks to BabySeq, parents are provided with early warnings that help them prepare for future disorders in their children. As you might expect, this project has been (and still is) controversial. Opponents are alarmed by the violation of children's rights because parents are sequencing their DNA without the child's consent. The debate is open.[123]

---

[123] (Genomes 2 People, 2018)

ii. Technology vs. doctors

To begin with

Medicine has tremendously increased our life expectancy. Since 1900, the global average life expectancy has more than doubled and is now approaching 70 years (in developed countries, the difference is even higher).[124] This is not only because of the availability of better food and drinking water but also because of the development of medicine and better access to it. Doctors and drugs have become key to increase our survival chances. But is there a limit to human life expectancy? Is there a roof? Do you also want to live… forever? Or even better: do you also want to stay… forever young?

Artificial intelligence for decision making

Wikipedia is probably one of the most popular websites on the internet. It is limited, but its aim to gather all the knowledge of humanity in one place is humbling. This idea is far from being new. Libraries, theaters, circuses, schools, universities, and forums have all been, and still are, magical spots to gather and share knowledge. Wikipedia is just the latest version of that dream. But having Wikipedia available on your browser doesn't make you a more intelligent person. Why? Because you cannot process all this information at the same time fast enough. What if you could not only gather a lot of know-how (far more than a human brain can store) and also process it to provide insights?

Imagine that you could gather and monitor all the medical indicators from thousands of people worldwide over several years. This information would include the diseases they had, the treatments they received, and the outcomes of those treatments. A good cardiac surgeon is a person who has operated on several thousand hearts in his professional lifespan and has the ability to predict what the best cure for your specific problem is. But his knowledge is limited compared to the potential knowledge of a machine that could correlate your pathology with every person

---

[124] (Roser, 2018)

who has had the same symptoms as you. And imagine that this machine could recommend a personal and unique treatment specially conceived for the characteristics of your heart, your lifestyle, and your genetic composition. But it doesn't stop there! With some medical indicators, genetic information, and big data processing, you could forecast health risks and work to reduce them even before the problem appears. You would be able to live longer, stay healthier, and be happier.

Let me tell you something: we are not talking about the future of health. We are talking about the present. Companies like IBM, with Watson Health, are moving toward a more integrative way of using historical health data. And considering the ever-increasing aging population, it has the business potential of being worth billions of dollars.[125] Healthcare as a whole is only beginning to see the potential of predictive analytics. In the next few years, the power of predictive algorithms will increase simply because there will be more data and technology available. Modern clinical datasets and analytic technologies present an enormous opportunity for humankind to further increase our life expectancy. The predictive power of models developed on tens of millions of individuals exceeds that of any traditional method. Humans have limited memory; machines have unlimited memory with almost infinite processing power. Predictive models will provide the best data-driven approach to a patient's health, helping your doctor to treat you better.[126]

Keeping track of yourself

Just as we saw in the rise of automation, the key element of this revolution is sensors. We will not only put sensors in our factories but also on our bodies, to track everything you can possibly imagine. This information will be essential for our doctors as well as ourselves. The most used tracking tool by far is our smartphone. Almost all new smartphones are pre-configured to

---

[125] Today, 8.5% of people worldwide (617 million) are aged 65 and over and this percentage is projected to jump to nearly 17% of the world's population by 2050 (1.6 billion). (National Institutes of Health, 2016)
[126] (IBM, 2017)

check your physical activity by counting the number of steps you take per day. But the potential of this device alone is much more significant. Just by installing some free apps, you will be able to track your sleep, your physical activity, your breathing rhythm, your diet, etc. This potential is further amplified when your phone is combined with wearables.

Current wearables can measure almost any health parameter you can imagine. For example, Google has developed contact lenses that measure the blood sugar level in tears to help people with diabetes. AliveCor has developed a device smaller than a credit card that captures your electrocardiogram to prevent strokes. Withings sells a portable, wireless device that monitors your blood pressure, while MicroCHIPS directly injects a chip under the skin below your waistline to deliver regular doses of drugs that are traditionally delivered by injections. In the future, you may be able to inject yourself with devices and sensors that you can carry 24/7. Information collected from these wearable devices will then be shared with your doctor to provide an accurate overall picture of who you are and what your lifestyle is. Your doctor will therefore be able to give you a better prescription that is specially designed for you. It is definitely better than asking you how many times you brush your teeth!

Robot-assisted surgery

In addition to an extensive database of worldwide patients and access to your particular and unique information, your doctor will also have the best tools to assist you, even wirelessly! Welcome to the era of robot-assisted surgery.

Today, one of the main problems in the medical field is that doctors are not accessible anytime, anywhere. There are some countries of the world that have great experts in one specific medical domain but lack experts in others. Moreover, the best doctors are usually located in big cities and not in small, remote areas. Therefore, if you need an urgent operation but the specialist you require is not available, you may have to wait for months. These are old problems that could potentially be solved by technology.

Imagine that the world had a global pool of doctors that were available for you. You would only have to go to the main hospital of your city and get the surgery done. Because your full medical data would be sent beforehand, the doctor would understand your disease and know precisely what your needs are (even if the doctor doesn't speak your language or even know you!) An economy of this scale will ensure that more doctors are available globally and will increase the efficiency of the overall system. And the quality of the operations will also improve; a robot is never tired, is never nervous, and has extremely high precision. Of course, robotic surgeons probably won't go mainstream for another 10 years. But this type of solution can also be beneficial in times of war and/or in developing countries where there is almost no access to medicine. Organizations like Doctors Without Borders already have projects that send medical help to countries in need. Imagine that they could go to the next level and have a fully ready surgical ambulance to send to hot zones. The ambulance would be available 24/7, and always staffed due to the rotations of doctors. The possibilities are amazing and endless.

Dear reader, don't think that we are talking about the future. At the time of writing this, robots like the Da Vinci Surgical System have already performed minimally invasive surgery on more than 3 million patients.[127] Impressive.

Final thoughts

To sum up, the advances in technology will completely change our perception of the role of a doctor and the place that medicine has in our lives. Technology will simply and naturally be part of our bodies and the data we generate will be shared with the people who matter to us. By combining all this data together, we will have an overall picture of who we are and how we can

---

[127] (Da Vinci Surgery, 2017)

better solve our health issues. Technology will serve as a vital tool for our doctors and will give them the chance to truly understand us—even better than we sometimes understand ourselves.

Box 1: Watson and cognitive computing

IBM's vision is that cognitive computing can help health providers with data integration, analytics, and coordination of care in a comprehensive manner. "Cognitive computing" means that Watson (the computer from IBM) can think like a human. It can analyze and interpret all of your data (including unstructured text, images, audio, and video) and provide personalized recommendations by understanding your personality, tone, and emotion. It can learn from each patient and apply its expertise to new patients. Watson can also interact with patients through bots that can engage in realistic dialogue. Beyond scanning health records, Watson is able to identify chronically ill individuals who may be at risk of a health emergency, help align care plans with data, help users execute their care regime obligations, and support clinicians to track compliance. Watson wants to become your virtual doctor, being able to *understand* you and your feelings. Being, somehow, able to *understand* you as a human. [128]

Box 2: HumMod and the end of human experiments

HumMod defines itself as "the best, most complete, mathematical model of human physiology ever created." Their ambition is to end all experimentation that is currently performed on humans (or at least to reduce the need for it as much as possible). With more than 10,000 variables, the HumMod mathematical model can create a whole human body with all its parts interconnected. Like we discussed in our genetic engineering section, HumMod gives you a "digital twin" to play with. Scientists can use these "twins" to test the effect of specific drugs on your body. By doing so, scientists can reduce costs, time, and human risk, thereby increasing the overall efficiency of their research. Without a

---

[128] (IBM, 2016)

doubt, HumMod is a promising approach to human experimentation.[129]

Box 3: eCare21 and the future of monitoring

eCare21 is a remote patient-monitoring system that can collect thousands of pieces of health data every minute of the day. It uses smartphones, Fitbits, Bluetooth, and sensors to gather information about everything that can be monitored: blood pressure, physical activity, glucose levels, medication intake, weight, and more. All this information is then compiled in the cloud and presented on a dashboard so that you, or whoever you want (doctor, family members, caregivers, etc.), can keep an eye on it and provide proactive care. It is expected that by 2020 the global telemedicine market will reach more than $34 billion—a trend to watch out for.[130]

---

[129] (Hummod, 2017)
[130] (Forbes, 2016)

iii. Becoming your own doctor
To begin with

Despite all the advances that we have presented for the years to come, a critical question remains unanswered: will I need a doctor to cure me and take care of me? Some of the responsibilities of doctors have been *outsourced* to technology, and the trend is quickly gaining weight. Your doctor will soon become the person that you will reach out to only in case of emergency. In the long term, in a fully monitored world, you may never have to meet a doctor in your life; you will simply never need one. It may sound crazy, but think of retail banks. They are already disappearing, little by little, due to the eruption of fintech companies. Will we see the same trend in the medical industry?

Social media and health

It is a fact that social media provides patients an opportunity to easily interact with physicians, nurses, and other patients. There are a lot of blogging sites that offer patients the chance to express opinions and share their experiences. These blogs often provide patients with an excellent source of information about a particular physician, hospital, or medical procedure. Moreover, these blogs comfort patients, making them realize that they are not alone in their health battles. Social media has been an excellent source of confidence, especially for patients who have long-term diseases. Twitter allows patients to interact and discuss conditions and experiences in "real time" and give each other personalized advice. These debates and discussions lead to better patient engagement and higher hopes. When patients are involved, they are more aware of their disease and treatment plan, are more likely to change their lifestyle, and are more compliant with taking their medicines. Knowing that other people have the same disease gives patients the emotional support they need to fight it. This enhances cooperation and reduces the support needed from psychologists and doctors in general, thereby lowering the high costs of medical services.[131]

Moreover, information on social media can have a direct influence on a patient's decision to seek a second opinion, especially those coping with a chronic condition. Patients and caregivers will always seek out second opinions, and searching the web can help them discover potential alternatives to diagnoses and/or treatments they otherwise wouldn't know about. This challenges the medical community, as patients are better prepared and already know the basics of their condition. Sixty percent of consumers in the United States say they trust doctors' posts versus 36% who trust posts from a pharmaceutical firm.[132] We definitely trust social media more than we think we do.

But we are not stopping here. We are witnessing a change in our social health habits. This trend is especially intense for the newer generations: 18- to 24-year-olds are more than twice as likely to use social media for health-related discussions than 45- to 54-year-olds.[133] However, it is a myth that only young people want to use digital services. Patients from all age groups are more than willing to use digital services for healthcare, with more than 75% of all patients expecting to use digital services in the future.[134] We are witnessing a change in the way we interact with other patients and doctors, making the overall community more aware of the realities of the diseases we suffer. This trend is expected to grow, which bodes well for all involved stakeholders.

Wearables will help you to know your body

The wearable technology market grew 29% in 2016, with 101.9 million units sold, and it is projected that it will reach more than 213 billion units by 2020. These new devices that are connected

---

[131] (Forbes, 2016)
[132] Devil speaking: this reality is of course exploited by the large pharma companies that *sponsor* influential doctors to post about their drugs.
[133] (ReferralMD, 2017)
[134] (McKinsey & Company, 2014)

to the internet even have a specific name: the IOHT or Internet Of Health Things. IOHT devices can provide timely updates to you and your physician about various health parameters, ranging from how long you sleep to how many steps you take each day. There are four categories of IOHT devices that can drastically change our health habits:

Fitness wearables: These are the most popular devices. They track fitness-related metrics such as distance walked or run, calories consumed, and, in some cases, heartbeat rate and quality of sleep. These trackers are synced wirelessly to your smartphone and can help you with your athletic goals and weight loss objectives. Or they can be just for fun!

Real-time health monitoring: These devices can provide the latest healthcare information with two interfaces: one for patients and another for doctors. In a simple example, let's take the case of remote cardiac patients. A real-time wearable device can detect underlying heart conditions which might help in preventing possible heart diseases or in recovering from cardiac diseases using parameters such as heart rate, blood pressure, and body and skin temperature. This information can help the doctor to diagnose and quickly detect conditions such as arrhythmia, hypotension, hypertension, and hyperthermia through an alarm system based on upper and lower threshold values.

Chronic disease prevention: In the United States, an astonishing 85% of the healthcare budget is spent on caring for chronic diseases, which are the most common, costly, and preventable of all health problems. Some examples of chronic diseases include diabetes, asthma, heart disease, and chronic pain. According to the Center for Disease Control (CDC), there are around 44,000 asthma attacks every year in the United States. These attacks happen because there is a lack of awareness about the warning symptoms that occur before an asthma attack. There are solutions on the market that could help fight diseases like asthma. The ADAMM (Automated Device for Asthma Monitoring and Management) is an intelligent wearable system that is connected to your smartphone. The device is attached to the upper torso and tracks warning signs of asthma attacks, including respiration pattern, cough rate, heartbeat, and body temperature. The device is then able to warn the patient of an imminent attack and make recommendations about what to do.

Wearables for eldercare: Generally, older people are more prone to health problems compared to other groups, and 22% have at least one limitation in vision, hearing, mobility, communication, cognition, or self-care that requires attention. Different smart wearables can help older people with their specific needs. For example, some devices can help connect them to relatives or emergency services in the event of an accident—even if they're unable to call for help themselves. Technology is developing a lot. For example, ActiveProtective has developed a wearable airbag that automatically deploys when the wearer suffers a fall, reducing the risk of serious injuries like breaking a hip. This is particularly important as the injuries can be difficult to recover from in later life.

With all the tracking devices that are currently available and yet to come, you will be able to track and improve your health without the help of any central public agency. You will know what you need to eat, how many hours of sleep your body demands, and when to stop running because your heart is beating too fast.[135] [136]

The internet as a tool to find answers

Last but not least, it is common knowledge that the internet knows everything. Information is easy to access, free, and fast. Therefore, when you start to feel ill, the first thing you do is check the symptoms on Google and then try to get an initial diagnosis and suggestion for a cure. If you don't feel terrible, you talk to a pharmacist to verify what you read online and buy the medicine that you saw online (if you haven't already purchased it through Amazon's one-day delivery service).

---

[135] (Khokale, 2017)
[136] (Techradar, 2016)

Technology can make this procedure even easier. Free apps like Symptomate Symptom Checker or free websites like WebMD (check Box 1) enable you to auto-assess yourself quickly and easily. The numbers are impressive. Roughly 1% of the searches on Google (think: millions!) are symptom-related.[137] The app world is also hitting a boom; mHealth was worth $23 billion in 2017 and is estimated to grow more than 35% annually over the next three years.[138] This trend comes from the fact that the largest generational segment in the western world (the Baby Boomers) are beginning to retire and are becoming familiar with technology.[139]

It is a fact that health websites have saved millions of dollars for insurance companies and governments by helping patients so they don't have to see a doctor. But it is also true that some patients have misdiagnosed themselves and have had complications as a result. Still, because we are moving toward a more *want-to-have-it-now* society, the overall experience of going to a hospital to see a doctor is losing appeal, especially for the younger generation. If societies are well educated about the tools the internet can offer, the potential of the move toward virtual healthcare is huge (in terms of cost savings and efficiency). However, there will always be moments when it is necessary to contact a specialist to solve your health problem. Non-digital channels will continue to be relevant and important, so digital channels will have to be embedded in a thoughtful multichannel design.[140] The internet is changing the way we access health services and can become a great source of solutions if used with good judgment.

Final thoughts

In brief, even if doctors are still central to the current health system, some of their functions are being replaced by technology. The fully-monitored health system is still far away.

---

[137] (Google, 2016)
[138] (Reuters, 2017)
[139] (ReferralMD, 2017)
[140] (McKinsey & Company, 2014)

However, the arrival of new technologies is saving, and will save, billions of dollars, making the health system more sustainable in the long term. As an estimation, using 2011 as a baseline, if these technologies were scaled up to create a system-wide impact, the savings could account for $300 billion to $450 billion in reduced healthcare spending. That's 12 to 17% of the $2.6 trillion baseline in US healthcare costs. A similar path could be expected in other developed countries.[141] Healthcare trends are changing, and technology is what's making the changes possible. In the future, we will not only be able to access health advice easily, we will be part of an ecosystem that encourages us to follow medical advice and makes us feel less lonely in our weak moments.

Box 1: WebMD as a symptom checker

WebMD wants to fulfill the promise of providing comprehensive health information on the internet. They want to be the Wikipedia of health. They provide credible information, supportive communities, and in-depth reference material about the health topics that might matter to their users. My favorite part of this website is the symptoms checker. By answering some questions on a smooth app, you can assess yourself and get an initial quick diagnosis. The app is free and user-friendly. It adequately represents the new paradigm of the twenty-first century.[142]

Box 2: Ginger.io and the future of self-tracking applications

Ginger.io is a mobile application that tracks behavioral health therapies through you smartphone. It records data about calls, texts, geo-localization, and even physical movements. The patients can also respond to surveys delivered over their smartphones. The cool part is that Ginger.io merges your data with public research on behavioral health from the National Institutes of Health. The reach of this app is impressive; it even offers unlimited chat with a coach, 24/7. For example, a lack of movement could mean that a patient is feeling physically unwell, while irregular sleep patterns may signal that an anxiety attack is

---

[141] (McKinsey & Company, 2013)
[142] (WebMD, 2017)

imminent. In the best case scenario, those risks can be assessed and counteracted before it is too late![143]

Box 3: Google Flu is faster than national health agencies

In 2008, Google launched Google Flu. This algorithm can detect an outbreak of influenza long before healthcare personnel can by simply compiling flu-related search words and tracking their entry into the search engine. In other words, by studying how many times people google their flu symptoms, the company can know where and when a new episode of flu is likely to be triggered, alert the health system of the country, and alert the public in general. This system could allow citizens to be more aware of the risks of a specific disease in real time and by location. Google Flu's predictions have generally been consistent with conventional surveillance data collected by health agencies, both nationally and regionally.[144] [145] This is an example of how technology could empower societies by providing real-time data about the spread of a viral disease, such as the flu.

---

[143] (Ginger.io, 2017)
[144] Google Flu Trends is no longer publishing real-time estimates due to privacy concerns. Historical estimates are still available for download, and current data is offered for declared research purposes.
[145] (Baoying Wang, 2015)

Genetics, automation of medicine, scientific discoveries, and self-tracking devices will not only change how we understand health but also how we understand life and death. Diseases will continue to decrease, and the future generations will grow stronger and live longer. The question is: do you know what you are going to do with the extra years you will have compared to your parents? Any cool projects in mind?

C. Will **you** become your own boss?

It is pretty cool how technology will shape every industry on the planet and even change our bodies. But the fun part is that it doesn't stop there. Technological changes will also have an impact on what we expect from society and work. It is time to take advantage of everything that technology can give you. Let's take a look!

i. It's time to break down barriers
To begin with

When you want to achieve something, there are always barriers. That's indeed part of the game (and part of the fun, too). The fact is that the new paradigm coming from this industrial revolution will remodel old barriers and build a new reality. The three main categories to take a look at are education, entrepreneurship, and globalization.

Education: free and accessible

A good education has always been easily attainable for the wealthy. Information is power. Literacy is a part of the United Nation's Sustainable Development Goal 4, which aims to "ensure inclusive and equitable quality education and promote lifelong learning opportunities for all." Indeed, significant progress has been made when it comes to accessing education. However, there are still 103 million people worldwide who lack necessary literacy skills, with more than 60% being women.[146]

The good news is that *cheap* internet access is changing the rules of the game. There are reasons to be optimistic about the future of education. Nowadays, information is available online. There are not many things that Google cannot help you with. How-to videos on every subject are becoming more and more popular, and the best schools are offering their courses online for free. (Just check websites like *Coursera* that have over a thousand courses from Yale, Stanford, and other prestigious universities).

---

[146] (United Nations, 2016)

The same goes for learning materials. Just to give you some examples, the Massachusetts Institute of Technology has made their course materials available on the web for everyone to freely access.[147] If you want to know more about the finite element analysis of solids and fluids, you merely have to search the website, download the material, and start studying. Do you want to learn how to code? *Code Academy* might be the place for you. Do you want to learn a new language? Sign up on *Duolingo* and join the former American president Barack Obama in his Spanish classes. There are thousands of websites that offer free and quality educational content. The impact of these tools is almost immeasurable. If you want to learn anything, you have the means. Never in history has it been so easy to educate yourself.

Moreover, new technologies and big data also allow for personalized learning. As we saw in past chapters, in the future, you will be able to design the car of your dreams and have it specially made for you. Well, the same applies to education. Your virtual and/or human teachers will analyze your learning process, move at the speed that suits you, and teach you the topics that interest you at the pace that fits you. You will be in the center of their attention, and no one will be left behind because we will finally realize that everyone is different. Cool, right? Moreover, by factually assessing your strengths and weaknesses, you will be able to better position your skills in the market. Your uniqueness will be understood and enhanced. Forget about the current system where kids of the same age study the same books and subjects. Everyone is unique and has their own personal dreams and passions. We are in a new paradigm where human potential is just waiting to be exploited.

Entrepreneurship has never been easier

Intrinsically connected to education, the eruption of the internet and of globalization has also created new entrepreneurial opportunities. Business barriers are now much easier to overcome. Entrepreneurial knowledge is far more reachable, and

---

[147] Check www.ocw.mit.edu

companies themselves are moving toward activities that require less capital and human labor.

Let me explain. Before, if you wanted to establish a new business such as a hotel, you needed at least an empty house compliant with the law of the area, a license, an advertising budget, some personnel, and a good working knowledge of taxes. Now, if you have a room in your apartment that you don't use, you only need 10 minutes of your time to open an account on Airbnb and upload some pictures of your room. *Et voilà*, you are done! You have built your small company! The barrier to entering the hotel business has decreased enormously!

Another example: Imagine that you have a great idea about, let's say, a new luxury watch that you believe others would really like. Before, if you didn't have money to start production and marketing, it would have been very difficult to shine in the market. Now, if you post your idea on sites like Kickstarter and garner support from the users of the platform, you can get some funds and compete with the big sharks of the luxury watch industry.[148]

Technology has leveled the playing field for outsourcing too. Through the internet and platforms like Alibaba, you can easily contact a T-shirt supplier in China to purchase 100 white T-shirts, then ship them to a factory in Turkey that will add your custom design. Cool, right?

Last example: imagine you have a skill that you believe can be useful to someone. You merely have to make a profile on a website like *Freelancer* and, if your skills and price match the demand, you can make a living. Easy, right?

Remember that if the barriers to enter a market decrease, the cost of failure is also reduced. It's becoming more comfortable to

---

[148] The example is based on the real story of Filippo Loreti, whose Kickstarter campaign managed to get more than 4 million euros from over 18,000 different supporters to start his luxury watch brand. Souce: kickstarter.com

figure out if your product is too pricey, if your customers are not yet ready, or if the idea is not good enough. The low risk of launching a new product allows entrepreneurs to pivot among ideas and business plans easily. Entrepreneurship is more comfortable and cheaper than ever before. You have an idea? The world is yours to make it happen.

But maybe you are not the entrepreneur in your family. You'll still be able to take full advantage of the power that automation will have over societies. Among other factors, rising incomes in emerging economies, coupled with increased spending on consumer goods, healthcare, and education, will create large-scale demand for new labor. This reality, combined with the healthcare needs of an aging population, real estate construction, and energy transition needs (from fossil fuel to renewable) will have a direct impact on the societal workforce. It is expected that between 555 million and 890 million new jobs will be created by the industrial revolution that is currently happening. Even more, we sometimes forget that new technologies also bring higher productivity. Therefore, companies can produce more at a lower price. Thus, the population can buy more things with less money, which simultaneously increases the number of things that we will consume. This increase in consumption will create new jobs in the middle term. Great news if you are looking for one![149] [150]

Globalization

Globalization has become a buzzword. Everyone talks about how cool trade is and how it will bring wealth to the world. International organizations like the United Nations, World Trade Organization, International Monetary Fund, etc., are working toward a more inclusive and open world–and market. There are several international economic unions[151], multilateral free trade

---

[149] (McKinsey & Company, 2017), the study was done in 47 countries.
[150] (Vox, 2017)
[151] Caribbean Single Market and Economy, European Union, Eurasian Economic Union, Mercosur, Gulf Cooperation Council, Central American Integration System, etc.

areas[152], and an infinite list of bilateral commercial agreements that make this world a smaller place. Policymakers are trying to promote these types of partnerships to boost exports through win-win contracts that enlarge domestic markets. At the end of the day, multinational companies are becoming prime examples of capitalistic success and are making headlines every day. There are enormous opportunities for companies to create value by taking full advantage of the falling barriers in regulation, economies of scale[153] and scope,[154] transportation, and infrastructure costs. Traditional national companies need to rethink their entire business processes with competition coming from everywhere. Being global is not *El Dorado* anymore; instead, it's the unavoidable Damocles sword to fight with. However, as a result of competition, liberalization, and new technologies, countries and companies can specialize in their best competitive advantages to battle around the globe with their products and services.

Globalization is even more essential for developing countries than for developed countries. Through the application of capital, technology, and know-how, multinational companies can create positive economic value in host countries and across different industries. Data shows that countries that open up to foreign investments enjoy improved standard of living, as consumers directly benefit from lower prices, higher quality, and a broader

---

[152] Central European Free Trade Agreement, Commonwealth of Independent States Free Trade Area, Common Market for Eastern and Southern Africa, European Free Trade Association, Pacific Alliance, Southern African Development Community, etc.

[153] Economies of scale: reductions in the average cost per unit coming from increasing the scale of production for a single product type. Let's take the example of a machine that costs $1,000 (fixed cost) and produces each unit at a cost of $1/unit (variable cost). If the machine produces 100 units, the total cost is $1,100. Therefore, each unit costs $11. If the engine produces 1,000 units, the total cost is $2,000, and each unit then costs $2.

[154] Economies of scope: reductions in the average total cost per unit coming from increasing the number of different goods produced. Let's take the example of two companies, one that sells apples and another that sells pears. Each company will need a marketing director, a supply chain manager, a sales person, etc. And all those costs will reflect on the price of the product. However, if those two companies merge, some positions can be filled in by a single person instead of two, thereby decreasing the average cost of each apple and pear.

selection of goods. This improved productivity and output indirectly contributes to an increase in national income. Foreign direct investment is already having a dramatic impact on the way companies do business and the manner in which developing economies integrate with the global economy.[155]

Last but not least, it is essential to emphasize the crucial role of cheaper transportation for business and globalization in general. Transportation is no longer a barrier for exports, primarily through the sea. Taking into account that ocean water covers 70% of the earth's surface, if you have a port on your coast, you can easily access numerous countries and their markets. But maritime transportation is not the only thing changing the shape of global business; the airplane industry is also taking a prominent place in this revolution. Currently, some low-cost airlines offer incredibly cheap flights[156] that ease business and tourism among countries. In this age, being international is the new normal even if it brings massive challenges and uncertainties with it.[157]

Final thoughts

The industrial revolution that we are currently living in is making the world smaller. We share a connected world, and the trend is going to move toward more connection. You can easily interact with people overseas and start a joint venture with them, perhaps using skills you learned online or taking advantage of all the facilities that policymakers have prepared for you. Of course, business is never easy. But the degree of difficulty and the barriers you will face are far lower than they used to be. Ready to become the next Elon Musk?

Box 1: China's Belt and Road Initiative

China wants to become the leading actor in the world's fate and is working hard to make it happen. The Silk Road Economic Belt

---

[155] (McKinsey & Company, 2003)
[156] Thank you Ryanair for disrupting the industry and making every airline follow.
[157] (McKinsey & Company, 2012)

and the 21st-century Maritime Silk Road (aka The Belt and Road Initiative) is a development strategy proposed by the Chinese government that aims to "enhance regional connectivity and embrace a brighter future."[158] This initiative was first introduced by Mr. Xi Jinping, general secretary of the Communist Party of China, who wants to construct a large unified market broader than the borders of China itself. The objective is to expand both international and domestic markets through cultural exchange and integration. In other words, the Chinese government (through the Asian Infrastructure Investment Bank and the Silk Road Fund, both state-owned) is investing heavily in ports, roads, and railways. China wants to have strong neighbors that will then be able to buy and trade with Chinese products. China's *neighboring countries* (the project includes countries ranging from Ethiopia to Hungary) have signed long-term agreements to enhance this collaboration. The program is an estimated $5 trillion infrastructure spending spree that spans more than 60 countries across Asia, the Middle East, Europe, and Africa. President Xi wants to expand the influence of his country and make China the leading trading power of the world, and he is aware that economic success depends heavily on robust trading capabilities. The Belt and Road Initiative is the clearest example that the world is moving toward erasing economic barriers to make global trade happen.[159] [160]

Box 2: More nanodegrees and nanojobs

People are encouraged to finish a degree in order to land a better job. The deal is that the degrees or certificates you hold will qualify you for a specific position in a workplace. But the majority of degrees don't highlight specific skills.

To address this concern, companies such as Udacity are beginning to offer nanodegrees and facilitate digital badges. A nanodegree is an online education service that is changing the

---

[158] (The State Council of the People's Republic of China, 2015)
[159] (Quartz, 2017)
[160] (The Economist, 2018)

way adults get educated. The process is hands-on, project-based, and career-focused. Unlike standard educational systems, nanodegrees have highly specialized objectives. This new way of understanding education is currently receiving huge investments from companies like AT&T, Google, and Facebook, and seems to represent the future of education. Nanodegrees are much cheaper than traditional ones and share the potential of landing a job. The difference is that nanodegrees focus on specific skills rather than taking on general subjects that will not be needed in the user's dream job.

Nanodegrees are still in their early days, but the trend is continuously evolving with an increasing number of students enrolling. In fact, employers are more and more aware of the existence of nanodegrees and are starting to see their value. They are even asking their future employees for the specific skills that their industry needs at that particular moment. It is a way of staying up to date in a fast-moving society.[161]

Box 3: WeWork and the access to offices

WeWork is a company that provides conveniently located offices in San Francisco. The business has grown into a $16 billion company just by providing services to people. Presently, WeWork has over 64,000 clients who are using 83 work spaces in 12 countries around the globe.

So what exactly does WeWork do for companies? It takes care of all their office needs so they can focus on growth. Start-ups and nascent companies are classic WeWork clients. It's not easy to find a good office if you don't have the economic scale to afford it. Rentals, décor, and legal documents need to be taken care of before you can start operating. This is where WeWork comes in. The company started by renting a large building with good spaces in the heart of San Francisco and subleasing them to start-ups at a lower cost. A WeWork membership gets the start-ups access to spaces where they can use working tables and free

---

[161] (Upwork, 2015)

WiFi. If a team needs privacy, they can also rent a private office by paying a premium. Of course, a common area with a relaxing ambience and ping-pong tables are also provided. What's more, every floor has a beer tap that offers different types of beer brewed locally and internationally. Members also get access to the kitchen, where they can personally prepare their meal or coffee. Every day, various events are held at WeWork, from lectures and meetings to conferences and launches. Companies such as WeWork have decreased the costs for start-ups to lease office space and have conveniently eased the first steps for young businesses.[162]

---

[162] (Business Insider, 2016)

ii. A smart and connected economy

To begin with

Globalization is about connecting the dots. This industrial revolution is rapidly moving toward a more connected economy. Currently, companies are not just interested in their specific products; they also offer innovative services to complement their products. And it is through this connection that real value appears and the Internet of Things is created. The real added benefit comes from the network that connects the dots. Let's take a look at how smart homes will look, how they will create smart cities, and how the sharing economy will play a vital role in the marketplace.

Smart homes

The global market for smart[163] homes was worth $14.7 billion in 2017, with North America representing 48% of global revenues. It is expected that by 2025, 10% of households worldwide will be smart homes.[164]

Homes today are very different from what they were a century ago. The way we live and behave at home has also changed, with new hassle-free technology that has made our life easier. Automatic lighting, remote controlled appliances, fridges that order milk, or Alexa answering your questions are just some examples. However, there is much more. Smart homes also offer possibilities for energy and cost savings, greater home efficiency through automation, and improved home security. Four of the consumers' top choices for smart home and wearable device features were energy saving thermostats, connected lighting, auto-adjusting thermostats, and connected energy tracking. As an example, the United States Environmental Protection Agency found consumers could reduce energy usage by 10-30% using

---

[163] "Smart" means that the house has an intelligent home system that uses a smartphone app or a web portal as a user interface. Therefore, the houses that have devices that are controlled by automats such as switches, timers, sensors, and remote controls are not necessarily smart by our definition.
[164] (IHS Markit, 2017)

schedules and temperature settings of programmable thermostats. Smart homes have the potential to fulfill consumers' growing expectations of convenience, sustainable living, safety, and security.

There are still major concerns about the privacy and reliability of these smart systems. However, the future looks bright. Nest (acquired by Google), Amazon, Honeywell, Xiaomi, and Netgear are the main players in the industry, with the tech giant Apple following in their steps (with the HomePod and the acquisition of Ring). They all want to enter your home and connect it to their clouds; to, again, create a network of things that will be able to talk to each other. The key word here is "network." The challenge will be to make different products from different providers talk to each other using the same interface. The buzz words "Internet of Things" will have to manage this cross-talk and will become a reality in the near future, making houses conveniently follow your orders.[165]

Smart cities

The Internet of Things will have a broader impact than the walls of your home. With 60% of the world's population expected to be living in urban areas by 2030, cities will become one of the key stakeholders of globalization. Worldwide, more than 180,000 people move to live in a city every day.[166] Just to have an idea about its global impact, cities currently consume 75% of natural resources worldwide, produce 50% of global waste, and emit between 60-80% of the greenhouse gas emissions. In the areas of energy, waste management, environment, and healthcare, connected devices can help increase resource productivity while saving capital and time.

Smart city services and solutions can help governments achieve some of their core objectives such as minimizing waste, improving resource usage and transportation systems,

---

[165] (Atlantic Council, 2016)
[166] (Philips, 2015)

controlling electrical grids more efficiently, etc. Let's describe a classic city problem. An average American owning a car spends 101 minutes in it per day, with 15% of the time spent in traffic and 20% looking for a parking space. In other words, only 66 minutes per day are actually spent in "useful" driving. Improving traffic flows and introducing intelligent parking systems will reduce congestion, and therefore emissions, from "useless" driving time. One approach is using "color coding" systems, where private vehicles are coded according to their plate numbers. The cars whose plate numbers end in an odd digit can only enter the city on Mondays, Wednesdays, and Fridays. Cars with plate numbers ending in an even digit are only allowed during Tuesdays, Thursdays, and Saturdays. This system will have an impact on the quality of the city's air, fuel consumption, public transportation usage, CO_2 emissions, etc. However, it is clear that this system also requires a "smart" city with connected devices and sensors. In this specific case, we will need cameras along the main roads of the city to record the plate numbers of cars, a centralized system that will gather and sort all the data, and an automatic process to send fines to the cars that entered the metropolitan area on a forbidden day. This will transform cities into connected ecosystems. To a large extent, these changes will happen through new B2B or B2C solutions. However, governments (especially regional ones) can play a crucial role by providing a long-term perspective and investing in large-scale solutions like public-private partnerships. It is expected that the global market for smart city solutions and the additional services required to deploy them will amount to over $400 billion by 2020.[167] [168]

Circular and sharing economy

Circular economy is re-thinking progress. It represents an alternative to the classic *linear* economy, which follows a straight path of making, using, and disposing products. The idea is to try to use any resource as long as possible, extracting the maximum value from it and reusing it at the end of its lifetime. Today,

---

[167] (Bolívar, 2017)
[168] (World Economic Forum, 2015)

there's a world of opportunity to re-think and re-design the things we make so they will be re-usable. Through a change in perspective, we can re-design the way our economy works, by designing products that are "made to be made again." This philosophy has huge implications for the global economy, climate change, and our final perception of production chains.[169]

The circular economy can be applied to many aspects of our lives. However, its principles have the most significant potential in the world of manufacturing. On average, today's manufacturing facilities operate at 20% below capacity. In an advanced economy scenario made by the EU, the union would save almost $630 billion by combining three primary drivers. They include:

- extending the lifespan of a product (by predictive maintenance)

- increasing utilization (by reducing unplanned downtime and increasing overall equipment effectiveness)

- looping or cascading the asset (by improving information about the condition and using history of individual components)

---

[169] Everyone talks about this circular economy thing. But do you know what it is? Simply put, there are three key principles that describe it:
- Principle 1: "Preserve and enhance natural capital by controlling finite stocks and balancing renewable resource flows." In other words, the Earth has some resources that are finite (such as oil) and some that are almost infinite (such as solar energy or wind). Classic example: replace the production of electricity sourced from coal (finite) to solar panels (infinite).
- Principle 2: "Optimize resource yields by circulating products, components, and materials in use at the highest utility at all times in both technical and biological cycles." In other words, extend the use-cycle of a product. For example, long lasting blades should replace classic shaving blades that have a very short use-cycle.
- Principle 3: "Foster system effectiveness by revealing and designing out negative externalities, such as water, air, soil, and noise pollution; climate change; toxins; congestion; and negative health effects related to resource use." In other words, make production processes more efficient by reusing and recycling the waste produced. For example, raising cattle produces enormous amounts of methane (which is a potent greenhouse gas). One way of reducing the impact of this activity would be to recycle this gas into biogas that can be used to generate heat or electricity.

To be fair, reverse logistics[170] and remanufacturing are also subject to several risks such as fluctuating supply and demand or the compromised quality of products after they are remanufactured. Companies will have to be flexible enough (and that's where technology can play a crucial role) to decide what to do with each product. Resell the product for scrap value? Recycle it? Use the parts that are still in good condition? Assessing a combination of factors regarding the product's condition and the market for the product will be the key to success in this new circular economy paradigm.

But it doesn't stop there. Performance-based business models could also exhibit the power of a circular economy. In this model, a supplier retains ownership of the product, and the customer pays to use it. This provides the supplier with a fixed, contract revenue stream. Moreover, it reduces the total cost of an asset while simultaneously increasing profitability and the customer service value. The real magic of this model is that companies would be incentivized to produce lasting products; the longer it lasts, the longer they would retain the customer and the more profitable that customer would be!

Another form of the circular economy would be to use data to redefine the design of the product itself. The explosion of data sensors plays a crucial role. Imagine a gas-turbine manufacturer. The aggregation of real-time engine condition data provides insights into more productive, durable, and long-lasting engine design opportunities for the future. Simply put, the data of the past will inform the production process of the future generation of engines. [171] [172]

---

[170] The process of moving goods from their typical final destination to concentrate them at a central location, either to capture value (through reuse, remanufacturing, refurbishment, parts harvesting, or recycling) or for proper disposal.
[171] (World Economic Forum, 2015)
[172] (PwC, 2015)

Final thoughts

The solutions we have described will need to be driven by businesses and entrepreneurs forming synergistic ecosystems of services that will make the world a better place to live. Technology will increase asset and source productivity in both goods production and services. Everything is expected to move to a smarter way of producing, consuming, and acting. Policymakers will also be vital in making large-scale infrastructure investments, designing the regulations for new companies, and incentivizing important innovations. A smarter way of living will not only change how our house understands our preferred temperature, it will make us change our ideas about what a production chain should look like and what impact we should have on the Earth.

Box 1: DriveNow and forget about owning a car

Car-sharing will probably dominate the transportation sector in the years to come. Smart traffic solutions will reduce the costs for every driver and passenger and contribute to the fight against climate change.

DriveNow is a joint venture between BMW and Sixt that operates in large cities of Europe and the United States. (As I write this book, they are planning their expansion into new territories.) DriveNow is a free-floating service that allows vehicles to be picked up and left anywhere within a designated operating area. It has been successful so far, but it might not be the final solution.

The car-sharing market is facing increasing competition from other mobility services like bike-sharing and ride-hailing companies. Membership fell by about 4% (1.18 million people) in the United States between January 2014 and January 2015. This was the first time membership has fallen in the United States. There have been a few small successes with a station-to-station model, which requires a driver to drop the vehicle at a specific spot. However, it seems that the final business model has to be polished if companies like DriveNow want to dominate the market. The opportunities are enormous even for a heavily legislated industry like transportation. It will be interesting to see who wins the race and if DriveNow will be the transportation solution for tomorrow.[173]

Box 2: LeftoverSwap and the end of food wastage

The sharing economy is booming. Everything seems to be sharable with others: cars, energy, houses, clothes, etc. The purpose of the sharing economy is to conserve products and resources, recycle when possible, and minimize waste. Well, food is no exception. Not restaurant food like you find on Deliveroo or UberEats. We are talking about leftovers. Leftovers!

The app LeftoverSwap notifies you when there's leftover food in the area. When you receive a notification, you can go and pick the food up. It might sound strange or anti-hygienic, and it will definitely be a no-go for many. Moreover, it opens great challenges for regulators. Who will be responsible for food poisoning? The cook? The company? The user? A simple idea can become complex quickly.

But there's a real need for apps like LeftoverSwap. Developed countries are wasting up to 40% of their food. For example, British households throw away 20% of the food they buy. One possible solution for this real problem might be to simply share the leftovers. Even if it sounds like this app is designed for homeless people and for charity organizations, that's far from the case. Even professionals with more than enough money to buy food have registered on the app. Originally, the app's users were households, but membership has grown, and now there are even bakeries and restaurants on the app. Their unsold bread and food is given away for free, rather than it getting spoiled. People who enjoy food sharing have proudly declared that they have never received inedible food since the time they registered on LeftoverSwap. Only time will define its success.[174]

---

[173] (Fortune, 2015)
[174] (The Guardian, 2014)

Box 3: Fairphone as a viable business model

Fairphone was born to act against the mobile phone industry. It is well known that phones are produced to last two or three years on average, that they are built under unfair working conditions, and that they use minerals that come from shady sources. Worse still, the number of devices is massive, and they are rarely recycled. Today, there are more mobile devices than people in the world, and they are multiplying five times faster than we are![175]

Fairphone wants to change that trend and have a positive social and environmental impact from the beginning to the end of a phone's life cycle. They have based their production on four primary goals:

- long-lasting design
- fair materials
- good working conditions
- reuse-and-recycle friendly

Their values represent the circular economy principles, which many people hope will become standard in societies. This is a new way of producing and consuming goods. Fairphone's success will depend on the number of users they convince to buy into their principles, but it is definitely the beginning of a shift in the industry. They have also built a community around their project to make their voice heard. Join the hashtag #WeAreFairphone to learn more about them.[176]

---

[175] (Independent, 2014)
[176] (Fairphone, 2018)

iii. A new way of doing business

To begin with

Business is just the reflection of how society behaves. The jobs that we have today are different from the jobs we used to have. In the same manner, the jobs that we will have tomorrow will very likely be different from our current jobs. New jobs, new roles, and new ways of working are inevitable. Different generations have traditionally had divergent views on life and professional careers. Even if the following descriptions cannot be applied to everyone, they represent the trends in the job market.

Traditionalists (people older than 70) want to be rewarded for loyalty and experience with defined roles and responsibilities and with clear authority levels. Baby boomers (between 50 and 70 years old) believe in work-life balance, want to be rewarded for long hours, want the authority to make decisions, and value social responsibility. Generation X (35 to 50 years old) put the highest priority on work-life balance and value matrix organizations. They expect authority to make decisions and stay for five years on average at the same company. Generation Y, also called Millennials (20 to 35 years old), value empowerment, ownership, and self-development. They don't understand the concept of hierarchy. They believe in flat organizations and want, at some point, to start their own business. They stay less than three years on average with their employers. Finally, Generation Z or Digital Natives (younger than 20 years old) value remote work, rapid career progression, flexibility, and change. They expect 24/7 availability and speed in processes. They are expected to change jobs frequently and work with a per-project-basis mindset. [177] Let's look into how this new revolution will change our perception of life and our ways of working.

What people are looking for in organizations

In the past, we defined success in the job market as a person sitting in a cozy office of a famous company, riding in a luxurious

---

[177] (Unite, 2015)

car, wearing posh business attire, and seeing staff rush to meet the "boss" to discuss the day's meetings and engagements. In short, success was associated with a high-ranking position in a famous company. In the old times, career paths were always limited to the corporation itself. After earning a degree, you would join a corporation, stay there and get promoted from time to time till you reached the top—or retired. You spent most of your time stuck in a position until the next opportunity came along.

Today, success seems to be less connected to the company itself and more connected to the individual. People enjoy working freelance or on a per-project basis. They enjoy working without long-term bosses, exhibiting their skills on the World Wide Web and learning on-the-go from the new sources of information available to them. Technology has allowed this change. According to Intuit, contingent workers (part-time, contractors, temporary, and self-employed) are expected to increase in number by up to 40% in the US workforce by 2020. Already, there is a trend among young people to skip traditional career paths and become entrepreneurs. They want to set up their own businesses creatively. In the next few years, these people are expected to be their own bosses (if they are not already).

With this career scenario, the definition of success is still evolving. In the years to come, success will not only be defined by your monthly paycheck but also (perhaps even more so) by your sense of fulfillment in whatever you do. You could be helping people, providing services to others, inventing new machines, creating software and programs, or even just making other people happy. The definition of success is no longer universal and is now aligned with your understanding of the world itself.

Young people are primarily attracted to challenging tasks and opportunities. They look for organizations where they can express themselves, with values that are aligned to their own. They look for teams (and not HR departments) where they can find opportunities for growth and learning. That is why, most of them jump from one company to another every two or three years. They are seeking a fulfillment that might never be reached.

The most HR-advanced companies have acknowledged this trend among the young workforce. They have restructured career ladders as lattices. Instead of helping workers climb for a higher position in a straight line from the bottom to the top, they challenge workers to accept multiple roles, making them shine in varied tasks while growing with lateral movements within the organization. The idea behind this is to propose different paths to leadership positions. In this way, workers will not get bored, and they will be able to try different positions, allowing them to figure out their talents. Society plays a vital role in guiding today's generation. While companies are evolving in the way they hire employees and provide jobs, education should also be following the same path by emphasizing the importance of possessing multiple skills and not just good grades. Society will need to adapt its approach to education and its idea of intelligence if we want to outperform machines and find fulfillment in our work.[178]

Turning up the organizational chart

Organizational structure is considered the skeleton of an organization. Each bone in a skeleton has a function and the same can be said for each branch and level of an organization's structure. Each department and position has to ensure that the whole organization functions and moves toward its vision. The traditional organizational chart is a one-size-fits-all, top-to-bottom structure. It was considered to be the most efficient in the past, but in fast-changing environments, traditional organizational structures may not be as useful as they were before. And the market has changed.

The workplace has embraced fresh models of collaboration, openness, and information sharing. This evolution has been primarily driven by millennials who value autonomy and ownership of their work and expect their voices to be heard in the

---

[178] (Unite, 2015)

decision-making process of the company they are working for. Millennials were born in the computer era, which makes them comfortable with technology. Therefore, they expect organizational models to take advantage of the possibilities that computing power can offer. In the time of social media and instant gratification, it is essential to have fast organizational structures with only a few levels. This structure enables workers to adapt quickly to new environments and is flexible enough to make them comfortable with change. But it's a double-edged sword. Companies need to have a well-trained and tech-savvy workforce if they want to apply the technology that they need to outperform competitors. Rethinking the organizational structure will be a must for any company that wants to stay relevant in the market.

Future organizations are likely to be flattened, so everybody's voice is heard. The company can, therefore, make decisions quicker and innovate faster. Communication is the critical aspect of this strategy. The idea is that, by reaching the different layers of an organization more quickly, change will be applied faster and people will feel they own the innovations.

Every company has its own nuanced way of working, but from a theoretical point of view, there are five main structures. The reality of your company is, as you might expect, a mix of all of them:

Traditional hierarchy: Following the classic pyramid authority structures, traditional hierarchy represents a way of organizing power. Authority and orders move from the top to the bottom, and the different layers of the organization "filter" the decision-making process. Depending on how important the decision is, the level of seniority of the decision maker will be higher or lower. Roles are usually very defined and clear. The main setbacks of this system are that communication typically flows from the top to the bottom, which deters innovation, dampens engagement, and delays action. Usually there is a lot of bureaucracy, and decisions have some *lag* in their response to the market. The traditional hierarchy is not adapted to the current times of fast change and frequent disruption.

<u>Flatter organizations:</u> Inspired by the traditional hierarchy structures, a "flatter" structure seeks to widen communication and collaboration. It will remove hierarchy levels within the organization, reducing the number of layers between the CEO and the cleaning lady. Authority still rests at the top of the pyramid and still moves from the top to the bottom. However, there is a strong focus on teamwork, which challenges the classic status quo relationship between boss and employee. This is the most practical and scalable structure for large organizations. Also, it is the first step a company can take to adapting to current times.

<u>Flat organizations:</u> (sometimes called self-managed organizations): As the name implies, flat organizations are flat. There is one level of authority and one level only. Everyone has the same status, and there aren't different positions. A classic example of this type of organization is the software gaming company, Valve (see Box 2).

<u>Flatarchies:</u> They are a mix of flat organizations and hierarchies. They have a few hierarchical levels, but when you are in one specific layer, the authority is the same among all your colleagues. Flatarchies give freedom to employees to be innovative while keeping a certain level of vision and general forward thinking in the company. They can be applied to both large and small organizations, and they are useful when there is a need to develop a new product. The management might decide to isolate part of the workforce in a flat structure to combine the freedom and innovation from this type of environment with the vision to deliver the product that the market needs.

<u>Holacratic organizations:</u> This is the most radical organizational structure. It is usually more viable for small and middle-sized companies. Holacracy is based on decentralized management and organizational governance. Authority and decision-making powers are distributed throughout self-organizing teams rather

than being vested in a management hierarchy. Groups organize themselves by using regular tasks and attend governance meetings to identify the problems of the projects themselves. The idea is that employees will no longer be assigned to specific projects. They will check all the needs that each project has and choose to work on one of them. When the task is over, they will re-examine the needs of the projects available and will probably work for a different project. Employees will, therefore, choose projects that require their specific skill at that specific moment. The decision will be made based on their job role but not on their job title.

Each company will have to create the structure they want depending on their level of maturity, digital skills, and management vision. Depending on the uniqueness of the company, its sector of activity, and the type of competitors on the market, one kind of organization might fit better. However, moving from one structure to another takes time and can bring a lot of frustration at first. Still, it is the price to pay to be innovative and adaptive.[179] [180] [181]

It's all about purpose

No organization can be successful without defining a clear vision, mission, and life philosophy. In other words, every organization needs a clear sense of purpose. Only after they know their purpose will they become truly innovative and sustainable. According to the anthropology expert Simon Sinek, a purpose-driven organization is a group of people "who show up for the same reason, who work together to achieve something and will sacrifice so that the others may make it. Those are the organizations and the people that change the world."

The quest for purpose has always been part of our human nature. However, globalization, technology, the relatively peaceful period since the Cold War, new family structures, and the access

---

[179] (Unite, 2015)
[180] (Forbes, 2015)
[181] (Morgan, 2014)

to decent healthcare have all somehow contributed to a change in direction. We want the companies we work for to have a bigger purpose than profit. In the past, profit and shareholder value were the top priority among businesses. Today, savvy organizations also focus on their purpose, defining and advertising the reason for their existence rather than their strategies for making money. But that doesn't mean they don't make money. Survey results of the Deloitte Millennial survey and EIU Societal Purpose revealed that the success of a business should not be measured on profit alone. The study on meaningful brands asked a test group how much they agree with the idea that "profit is the sole basis of identifying success." The idea was rejected by 92% of millennials. In the current times and with the new expectations empowered by technology, it is important for organizations to impart their purpose to their employees, so employees too can know what they should be working for. An open dialogue should be held between employees and the organization's leadership. The main purpose of such dialogue is for companies to help employees align their personal aspirations with their work, to know whether they are focused on the right things, and to have a healthy debate before pushing toward new directions. Purpose, then, is considered a driving factor behind one's work, and its impact is definitely powerful and enabling.

Many organizations are now embedding "purpose" and "social good" in their performance reviews, training, and development programs. One example is Unilever's Performance Culture. It allows employees to have three business goals and one development goal which is relevant to the organization's sustainable living plan. Hence, the company has been rewarded by the impact they have on people. Leadership models too have changed their approach to cater to people's need. In London, "leadership laboratories" are emerging to provide support structures for coaching, consulting, networking, and other things that help individuals deliver profitable projects that create social impact. In Munich, the insurer Allianz has a development program that nurtures top employees, while addressing operational challenges in social enterprises. This company believes that companies need to understand the social dimensions of doing business. Through talent development, employees develop their leadership skills and have the opportunity to create innovative solutions to problems as well as finding new ways to do business. The Swiss company Credit Suisse has followed suit by setting up a new department to nurture socially conscious investing to empower positive impact. All these new models redefine the purpose of companies. The rules of the game have changed and are now more focused on sustainability and ecology than they were in the past. While increasing profits, employees of these companies also tend to find meaning in their jobs. If not, workers simply quit.[182] [183] [184] [185]

Final thoughts

My dad used to say, "Son, you have to work for the European Commission as a civil servant. Start as a young engineer advising policymakers, climb the ladder to the top, and in 20 years become a director of whatever." And I guessed he was right. Good money (by the way, not taxed), good career, low risks, good working hours... But things have changed. We are in a different generation, and we look for different things. Previous generations wanted a company to be loyal to. Forever. Well, at least until retirement. Now, people want a company with a purpose that is innovative and flexible. Forever. Well, at least until they change companies or die. But the younger generations are not looking for the type of "innovation" that every company on this planet seems to have on their annual reports. They are attracted by truly innovative culture. And if the company doesn't have it, they leave. No regrets. Nothing personal. It just wasn't a good match. They just swipe left. It's that simple.

If you check the new trends about doing business, you realize that technology will not just change the rules of the job market and how companies hire. It will also improve the overall concept

---

[182] (Unite, 2015)
[183] (Sinek, The Purpose Driven Organization w/ Simon Sinek, 2011)
[184] (Roser, 2018)
[185] (Reuters, 2017)

of work and professional relationships. Deals are not closed anymore through posh meetings; they are closed through WhatsApp. Big bosses don't wear ties anymore and sabbatical years have become the new to-do before turning 30. We expect the same things from our companies as we do from technology: speed, flexibility, convenience, and a willingness to listen. The question is:[186] is your company retaining the young talent they need to understand the new reality of the market, or are talented employees leaving after two years? Is your organization changing as fast as the market is? Will you be able to keep up in your industry?

Box 1: What it means to be Agile

Companies, in many different ways, strive to provide the best services they can offer to customers. Global institutions are therefore redesigning themselves to be more agile, aiming for both stability and flexibility. According to an analysis conducted in 2015, agile companies have a 70% chance of being highly ranked in organizational health, which means strong financial performance. Among 2,300 large companies in the United States, the research team identified 10 companies whose net income increased by at least 5% annually over 10 years. These high-performing companies were extremely stable. Moreover, the organizations featured rapid innovation and adjusted and readjusted resources quickly. The essence of true organizational agility is the ability to be both stable and dynamic. Agile companies design their organization with a framework of stable elements. These foundations are likely to endure over a reasonable period of time. It is important for an agile company to stay resilient and not destabilize even when unexpected challenges come along. Organizational structure, governance, and processes need to be balanced to achieve that goal:

- Traditional structures are based on boxes and lines on the organizational chart where specific tasks get done by workers

---

[186] You don't owe me a reply. It is not my business. But it might be time to have a grown-up conversation with your boss if you believe that your company is not ready for the market changes of this century.

with clearly defined responsibilities. Traditional structures usually have a boss at the top who manages and directs subordinates. An agile organization has a different structure. First, the management decides which dimension of this organizational structure will be the priority. Employees receive training and coaching according to where they are assigned. Awards and recognition are likely given to teams and not to individuals.

- Agile governance talks about situational decision making. There are decisions that are made by leaders alone. However, there are also decisions that require dialogue and collaboration. Some decisions are even made at the lowest position in the company. Leaders may delegate decisions to other people as much as possible.

- Processes determine how things are done. Companies should look into how they can uniquely (or at least better than competitors) deliver services or products to stakeholders. This is all about thinking up strategies to outperform the market in order to make the product grow as much as possible. Creativity is one of the essential tools in redefining processes.

For companies to become agile, they must rethink and, if necessary, redesign their structure, governance, and processes to strike a balance between flexibility and stability. The agile mindset might not be the best strategy for all the industries and roles. However, its principles can be applied (at least in essence) to the majority of cases.[187]

Box 2: Why Valve is unique and why I admire it

If you want your company to achieve greatness, take care of your employees. This is the only secret that Valve reveals in its handbook. Valve is a US-based video gaming company, founded by former Microsoft employees, that has built popular video games such as Half-Life or the platform Steam. Valve has become famous not just for its excellent products but also for its unique way of taking care of its employees. As a matter of fact, the

---

[187] (McKinsey & Company, 2015)

earnings per employee at Valve are higher than that at Google, Amazon, and even Microsoft. Valve claims that they are an entertainment company, a software company, a platform company, but most of all, a company full of passionate people who love the products they create. The most important characteristic of Valve is that, unlike other giant companies, it doesn't have managers. Employees are free to choose which project they wish to work on. Valve encourages employees to be responsible in whatever they are doing without anyone instructing them. By giving freedom, Valve creates the opportunity to liberate talents and trigger creativity among employees. Valve builds respect for the worker. The founders designed the company as a place where talented individuals are empowered to put their best work into the hands of millions of people without supervision. Valve also acknowledges the fact that failure is a part of success. Employees learn from their mistakes, and they are encouraged to fix failures, stand up, and start again. They are encouraged to try something new and discover more, removing any fear that may hinder their success. Aside from having no bosses, Valve provides amenities such as a gym, massage parlor, and full-service cleaning—so employees are free to relax, enjoy themselves, and have fun. Moreover, the care Valve gives its employees is extended to the employees' family too. This workplace might not fit everyone. There are still individuals who prefer traditional setups and prefer working with superiors in a more conventional approach. However, I recommend reading the "Handbook for new employees" from Valve. It is truly inspiring.[188] [189]

Box 3: The "job hopping generation"

According to a Gallup research poll, 60% of millennials are open to a new job and 21% of them have changed jobs within the past year. Gallup estimates that millennial turnover costs the US economy $30.5 billion annually. Millennials are by far the most educated working demographic in the history of developed

---

[188] (Forbes, 2012)
[189] (Valve, 2012)

countries. They want to be promoted fast because they believe they deserve it. The disappointing reality is that not everyone gets promoted and, even if they do get promoted, the company doesn't necessarily inspire them to dream. According to the National Occupational Employment and Wage Estimates in the United States,[190] there were 6,936,990 managers for 137,896,660 total employees in 2015. That means there was a ratio of nearly 19 employees to every manager on a national scale. In other words, only 1 out of 19 employees has the possibility of being promoted. The question is: what happens to the other 18? Well, they simply find their way inside the company through another department, or they move to another company. Gallup's report states that only 29% of millennials are truly engaged at work. The rest (71%!) are either not engaged (55%) or actively disengaged (16%). In other words, 7 out of 10 millennials would accept a good (or not-so-good) offer coming from the outside. Almost all the companies in the world are trying to stay innovative in this fast-changing world, and they will therefore need a young, tech-native workforce. Leaders are expected to understand how to attract millennials to their organizations and, once they are there, how to make them stay longer than their average two years.[191] [192] [193]

---

[190] Bureau of Labor Statistics
[191] (Recruiting.com, 2018)
[192] (Gallup, 2016)
[193] (Sinek, 2016)

Access to education, low business barriers, smart and connected societies, and new ways of doing business will not only change how we understand work and professional paths but will also improve our perception of authority and influence in an organization or society. Every piece of a company's puzzle will have to be heard in an efficient way, and technology will be the enabler of all these societal innovations. The question is: what place do you want to have in society? How will you take advantage of all the new opportunities on the market? Do you have any new businesses to start or any new skills to learn? The time is now, and luck is smiling on you. What are you waiting for?

We have seen that the industrial revolution we are living in is going to change our lives. A lot of revolutionary technologies are already on the market—not yet in critical mass but growing. And the neat thing about up-and-coming start-ups is that competition arises and, therefore, quality and diversity also move along.

Cool, right?

In time, industries will be able to deliver dream products designed by you for you. Science will make you and your kids stronger and better prepared for the dangers of this world. And society will create a flexible ecosystem to enhance your creativity and help you become a star.

And… well, that's all folks!

We've seen that innovation is speeding up. Speed and personalization are the new black. We will make all our desires come true.

Do you want to have it all? You can. **Period.**
Do you want to become Superman? You can. **Period.**
Do you want to be your own boss and work whenever you want? You can. **Period.**

"And they all found in technology the answers to fulfill their dreams and lived happily ever after."

But... isn't that too good to be true?

Of course it is, dear ;)

Y en las pistolas fíjense, a cada disparo recula el cañón como asustado por lo que acaba de hacer

- Xhellaz, Una mirada.

[Translation: "Check the guns. At every shot, the cannon retracts, as if frightened by what it has just done"

- Xhellaz, Una mirada.]

Order your favorite smoky and spicy whiskey on the rocks, and let the show begin ;)

PART
TWO

A new end?

Technology will bring us to the stars, but it will also make us realize that we didn't solve all the problems on Earth. It will strip us of everything we knew and took for granted. It will make us afraid of who we are and what we've done in our lifetime. And it will make us realize the following hard truth: we will be connected but, deep inside, feel utterly lonely.

A. Will **you** lose everything you know?

I don't like pessimistic people. I am not one of them. But I want to be realistic and honest about the current trends. Technology will ultimately shape how the job market and economy works. And believe me, there will be losers. Brace yourselves to hear some shocking and scary facts, but stay calm. The sooner we open a debate about the costs of development, the easier it will be to solve those issues and avoid disappointment.

i. The concept of work
To begin with

One of the indicators of a country's genuine development is its welfare infrastructure. Health and social security systems, education, and an attractive job market are intended to be the basis of what we call successful societies. There is a correlation between those factors and the overall success of the country. Politicians love to brag about the welfare that their countries can offer. However, data shows that welfare will soon need to be reformed and updated to match a new reality.

Jobs, life expectancy, and the automation of functions

This industrial revolution is extremely good for everything related to high speeds. Companies will need to be highly flexible to fulfill our ever-changing dreams. But costs will need to be paid.

One cost is that the speed of job creation (and destruction) will increase. We can forget about the *baby-boomers'* mentality: staying in the same company and job for their whole professional life. For one thing, the life expectancy of a company is decreasing. In other words, it is likely that your professional life will be longer than the life expectancy of the company you are currently working at. Fifty years ago, the life expectancy of a firm in the Fortune 500[194] was around 75 years. Now, it's less than 15

---

[194] The Fortune 500 is an annual list compiled and published by Fortune Magazine that ranks 500 of the largest US companies by total revenue. Important remark: they are classified by revenue only, not by profit or market capitalization.

years and declining even further.[195] For another, the needs of each company might change completely with time. A company that is 10 years old might switch from one business model to another completely different model. Companies are hiring less and less for indefinite jobs, and it makes sense in economic terms. Business models are changing fast and companies (and people) have to adapt to this change. The more variable the costs of the company are, the easier it will be to innovate and adapt.

The second significant cost of flexibility is automation and the famous *end* of human work. Let's be careful with those statements and take a step back. Machines are great and can do a huge number of cool things. But there are limits. At least for the near future.

A McKinsey report analyzed the current US workforce activities and looked into the percentage of time spent on tasks that could be automated by adapting currently demonstrated technology. table 05 shows the results of this analysis.

TASK	PERCENTAGE	SUSCEPTIBILITY OF BEING AUTOMATED
MANAGING OTHERS	7%	Least susceptible
APPLYING EXPERTISE	14%	
STAKEHOLDER INTERACTIONS	16%	Less susceptible
UNPREDICTABLE PHYSICAL WORK	12%	
DATA COLLECTION	17%	Highly susceptible
DATA PROCESSING	16%	
PREDICTABLE PHYSICAL WORK	16%	

TABLE 05: PERCENTAGE OF TIME PER TASK SPENT IN US JOBS AND TECHNICAL FEASIBILITY THAT THE WORK COULD BE DONE BY MACHINES.[196]

As the table shows, more than half of the tasks that US employees perform are highly susceptible to machine takeover. These tasks

---

[195] (Forbes, 2011)
[196] (McKinsey & Company, 2016)

are most prevalent in manufacturing, retail, hospitality industry, and food service. Still, it is important to remember that it's not only low-skill work that has the potential to be automated. Middle-skill and high-skill occupations are prone to a partial degree of automation too. As processes are transformed by the automation of individual activities, the workforce is predicted to perform activities that complement the work that machines do and vice versa. It will be a duo. Of course, the impact of automation will vary across different activities, countries, occupations, and wage and skill levels. Some studies suggest that half of today's work activities could be automated by 2055, but this could happen up to 20 years earlier or later depending on various factors, in addition to other economic conditions.[197] [198] There are indeed too many variables. But wait, let's go back to table 05. Would automation mean that half of the workforce is not needed? Hold your horses; there are several key points to consider:

This study takes into consideration the current jobs available in the US market. First, the United States is not representative of the whole world (even if it is a good estimation of the developed countries). And second, the study takes into account the *current* job market. Clearly, jobs will change in the next few years. An estimated 65% of children entering primary school today will ultimately end up working in completely new job types that don't yet exist.[199] [200]

A second factor to consider is, of course, the cost of hardware and software to develop and deploy new solutions. Technical feasibility is a must if you want to automate processes. But it is not

---

[197] (McKinsey & Company, 2016)

[198] (McKinsey & Company, 2017)

[199] (World Economic Forum, 2016)

[200] It is quite surprising that we have teachers and education systems that train the future employees of the world to do jobs that simply will not exist in the future. Like it or not, current education systems are not preparing societies to live in this new era. I simply wanted to raise some awareness.

the only requirement. The company has to see an exciting ROI[201] and payback[202] to automate a process.

We have seen that the facilities for outsourcing will notably increase. Therefore, with a higher supply of human workers, the price of a man-hour will decrease (by supply-and-demand dynamics). Consequently, it might become more interesting to outsource than to automate.

Last but not least, the social effect of replacing humans with machines has to be weighed too. First, jobs are socially protected (especially in European countries). It might be expensive for a company to fire someone even if it makes *economic sense*. Government regulations and social disapproval can't be avoided in the decision of automating processes. Second, some jobs have an intrinsic human need. For example, you expect a doctor to be human. Maybe a machine could predict and cure you much better and faster than a human person could. However, you expect human interaction for those types of jobs. The same applies to management. You expect your boss to coach you humanly, even if a computer could give you new and challenging KPIs specially calculated for your potential and motivation.[203]

Let me be clear with these points. Automation will hit the job market. It will destroy jobs, a lot of them. But the changes will be very diverse depending on the sector, the country, and the specific responsibility of the employee. As technology develops, robotics and machine learning will bring new techniques that will, for example, enable a better one-to-one between humans and machines. The trends show that the future of jobs is moving toward collaboration with machines rather than complete substitution of humans by machines. Those changes do not have

---

[201] ROI means <u>R</u>eturn <u>O</u>n <u>I</u>nvestment: how much you will get back from your investment. It must be cheaper to have a machine than to have an employee performing the task if you want to move toward automation.

[202] Payback is the time that it takes for an investment to be paid back. The lower, the better.

[203] (McKinsey & Company, 2016)

to be completely negative. They could also represent a great opportunity or a completely new way of thinking. However, here is where the hard questions of technology appear: until what point will you let technology enter and replace your current job activities? Until what point might your job be automated? Is your added value *human enough* to be irreplaceable? These questions are hard but necessary. You are the master of your future. It will be your task to find an answer and to adapt fast enough. Your job has a finite contract duration, so deal with it. ;)

Salaries and middle class, time to disappear?

First, let's set a basic framework. Supply and demand have an important influence in defining the price of an asset. This basic economic principle states that when supply increases and demand remains stable, the price will tend to decrease. Similarly, if demand increases and supply remains stable, price will tend to increase. An employee is, in economic terms, an asset like any other. The cost of hiring someone is also defined by supply and demand. This industrial revolution will indeed change the supply and demand on the job market, which will, by definition, change the salaries associated with any job. Of course, every industry/country/job position is different. However, there are general trends that I will explain. We will analyze this topic from two different perspectives: developed and developing countries. For each of them, we will take a closer look into the different layers of society: the blue collars, the middle class / white collar,[204] [205] and the *highly skilled* people.[206]

---

[204] The distinction between "blue collar" and "white collar" is based on occupational classifications that distinguish workers who perform manual labor from workers who perform professional jobs. Historically, blue-collar workers wore uniforms, usually blue, and worked in trade occupations. White-collar workers typically wore white, button-down shirts and worked in office settings. Other aspects that distinguish blue-collar and white-collar workers include earnings and education level.

[205] Disclaimer: every country and company in the world is unique. In the following lines, I will describe the average trends for the developing and developed countries. Take into account that each state may need a more in-depth personalized analysis to understand its concrete dynamics.

Let's begin with the developing countries. Their job market will entirely shift, with the population leaving the poor layers of society and moving toward a newly-formed middle class.

Historically, the majority of workers in <u>developing countries</u> have performed blue-collar tasks. The reason is that developed countries have been outsourcing their manufacturing to them. Developing countries have offered cheap labor and have authorized soft environmental legislation to take advantage of the ease of trade and low transportation costs. Unfortunately, many of the blue-collar jobs in developing countries might disappear as technology progresses and developed nations begin their own, automated manufacturing processes. For those countries that lack the infrastructure and know-how to adapt to the industrial revolution, things could get grim. Young unemployment might become the source of unrest and social instability.

On the other hand, the middle class that has access to the internet will have a range of new opportunities, mainly in the services sector. They will no longer be restricted to jobs in their local market; they can access jobs worldwide. Here is where globalization will present opportunities for the developing countries to succeed. Their primary asset is that they have a massive young population that is not afraid of new technologies and is aware of the culture of developed nations. This new middle class will bring internal consumerism that will feed economic development. In the long run, this new middle class will be the shining beacon of hope for developing societies.

Still, the lack of industrial infrastructure and highly-skilled workers might make it hard for these countries to fully take advantage of the industrial revolution. Of course, each state is different. But the challenges that they face are very similar.

---

[206] I define *highly skilled* people as those who have leading positions in companies (e.g., the management board) and / or the workers who have a particular skill that is vital for this industrial revolution (e.g., data analysts). Those positions are difficult to replace by machines (for now).

The key to their success will be:

- political stability, ease of doing business, low corruption, and robust legal frameworks to attract foreign capital
- education and infrastructure investments to set the foundation for development to happen
- development of a strong middle class and a decrease in poverty to retain those investments

Now, let's take a look at the <u>developed countries</u>. Even though the future seems brighter overall, even for developed countries, there will be winners and losers. Only genuinely innovative cultures will survive in this globalized world with fierce international competition. In the short term, there will be an expansion of the job demand. The factories of the future are yet to be built, and the know-how for it is precisely in the developed countries. Therefore, we will witness a curious phenomenon: the businesses that once left the country for developing countries will come back. Why? Because the know-how to build the fully automated society is more likely to be lodged in the brains of engineers from developed countries, thanks to their strong education systems and infrastructure. In general terms, before destroying jobs, automation creates them. In the middle term (between 5 and 25 years from now), the newly-built automated factories will reshape the job market, making blue-collar jobs unnecessary. However, the new technologies will also bring new opportunities.

Let's begin with the blue- and white-collar workers. As we have seen, in the short term, these workers will be much sought after to build the factories of the future. However, physical activities in highly structured and predictable environments, as well as data collection and processing, are the most susceptible to machine replacement. In the United States, these activities represent up to 51% of the economic activity, accounting for almost $2.7 trillion in wages.[207] There are two possible scenarios for their fate.

---

[207] (McKinsey & Company, 2017)

In scenario one, the middle class will join the new freelancing economy and become part of the low-middle class. Platforms such as Uber, Upwork, Freelancer.com, etc., will offer a source of income that was unthinkable before in their local markets. Here is where I see the higher costs of this new industrial revolution: we will have an incredibly flexible society that will allow us to switch jobs easily in a dynamic market, BUT we will have to give up part of our social protection and middle-class salaries and expectations. We will exchange money for flexibility and independence. In other words, we will earn less but have more freedom and less risks of being unemployed. Therefore, the middle class will decrease its overall wealth.

In scenario two, education will teach the middle class the new skills that will bring a unique added value to the world. In the following years, there will be hundreds of new jobs that did not exist before. If the middle-class society is able to learn these skills fast enough thanks to their quality universities, and capitalize on the first-mover advantage to be competitive in the globalized market, then the middle class will survive. However, this second scenario requires a culture of innovation and the willingness to be adaptable and do a different job every year. To be honest, if we want to maintain the middle class of developed countries, there isn't really a choice. Culture must shift to realize that borders do not exist anymore and that markets are not national anymore. Everything has become global. The speed of change is increasing; our response must be urgent.

Last but not least, the highly-skilled people will be more and more sought after by the companies that want to take part in this global industrial revolution. The potential returns for highly-skilled and more adaptable workers are quickly increasing.[208] The need for highly-skilled people will also become global. Different countries will compete against each other to attract and retain this talent that will be crucial in deciding if the nation will become a winner or a loser. There is indeed a bright future for highly-

---

[208] (UBS, 2016)

skilled workers, thanks to globalization and international trade, especially when talking about specific IT profiles such as data scientists or developers.

In my opinion, the most significant challenge for developed countries will be managing the middle class of a decreasing population, which will have to remain competitive in a globalized world. If middle-class jobs are not *human enough* or do not require specific difficult-to-find know-how, these jobs will be easily outsourced to other countries, to artificial intelligence, or to automation. Automation will continue to put downward pressure on the wages of low-skilled workers and is already starting to impinge on the employment prospects of moderately-skilled workers too.

As we have seen, there will be different movements in the job market for developing and developed countries. This industrial revolution will ultimately reshape the market at different speeds. Even if every state is different and has its unique traits depending on the type of industry/services it has, the question remains: where are you in your society? Are you one of those who will take advantage of the flows described? Will you stay where you are or make a move? Will your work be automated and forgotten? Would you prefer having complete freedom regarding your job or choose professional security instead? Are you working hard enough to be on the winner's side?

Politicians' nightmare, also called the future of public pensions

One of the most polarizing topics in our developed societies is the future of public pensions.[209] Health is and will get better—and that creates a problem. People live longer. If you check the statistics, you will find that we don't know what the cost of a retirement plan is. However, compared to life's other big

---

[209] This chapter will only refer to developed countries because the pensions industry is much more developed regarding revenues than in developing countries (even if there are notable exceptions). Also, the problem coming from the aging societies is mainly an issue for developed countries.

expenses (buying a home or raising a child), retirement carries the highest average price tag. If we take the example of the United States, the average cost of retirement is over $700,000 or about 2.5 times that of the average house. For many of us, retirement will probably be the most important financial decision we make in our lives. Let's take a look into the future.[210]

Here's a fact to put this issue into perspective: the percentage of people above 65 years of age in rich developed countries is growing as a result of rising life expectancy and lower fertility rates. It is expected that between now and 2050, the growth will jump from 16% to about 25% for the United Kingdom, from 12% to 21% for the United States, and an impressive 9% to 35% for South Korea. If we look at the 15 richest European countries and assume no policy changes, public spending on pensions will increase from 10.4% of the GDP today to 13.3% by 2050.[211] To be crystal clear, there aren't any magical solutions. The aging trend leaves only three possible choices: higher retirement ages (i.e., retire later), lower public pensions relative to average earnings (i.e., less income received when you are on a pension), or higher contribution rates (i.e., more taxes and/or more private pensions).[212] Of course, an excellent public pension strategy will be a combination of the three. However, those types of reforms are unpopular even if they are urgently needed. Currently, governments simply disguise the shortfalls in the pension system by using debt to cover deficits. This approach is unsustainable and will have to be stopped in the future.[213] But the social implications of the pension problem are even more important to consider. We have seen that this industrial revolution needs flexible, entrepreneurial, and tech-friendly workers. Older generations with 30 years of experience in a specific industry may

---

[210] (Merrill Lynch, 2017)

[211] At this point, you might think: 3% of the GDP of a country is not that much, right? Public pensions are funded by mandatory contributions from the workers. The GDP of a country is not a budget that governments have. For instance, the GDP of Germany was $4 trillion for 2016. However, the German government budget was $340 billion for the same year. This 3% is what will decide whether you retire at 65 or 80.

212 (McKinsey & Company, 2007)

213 (Rauh, 2017)

be faced with a dilemma. Their industry may be directly disrupted, or there may be new technologies that are shaping the market, making it difficult to follow. Generally speaking, 50- to 70-year-olds do not have the mindset for innovation and change. If retirement ages are prolonged, governments and companies might be compelled to retain people in the active job market who are not ready for this new era, making the overall society less competitive against other countries and companies. The second option of reducing the pensions seem unfeasible because this decision will make all the people who don't have a private pension unable to support themselves. The third and last option (higher contributions) seems to be the best solution to this problem—or at least the least controversial. There are several options: increase the contribution from companies/workers for the pensions, put new taxes on goods or services, give tax advantages for private funds to promote their use (leading to lower revenues for governments), create legal facilities for private funds, persuade people to save, make the pension system understandable and as simple as possible, increase early education on financial literacy (including at the primary school level), etc. There isn't a perfect answer to the pension's future crisis. Each government will need to take actions as soon as possible and open the needed debate on this delicate topic.[214] [215]

Indeed, the issue of pensions will be a hot potato in the upcoming years. This fixed expense will put government budgets under pressure, and new agreements will have to be made. When will the age-limit for workers to become pensioners move? What level of respect do our elderly deserve? Is it fair for the younger working generations to pay for a service that they might not receive in the future? Clearly, they will not have the same terms and conditions as their parents had. The questions remain: how comfortable are we with this fact? What is the right balance to strike and how sustainable is it? It seems that this issue is a sleeping beast that will soon awaken to shake the fundamental social peace between the older and younger generations. And

214 (World Economic Forum, 2014)
215 (World Bank, 2016)

when this unrest explodes (because it will), which side will you take? Are you saving enough?

Final thoughts

This industrial revolution will not only change the job market, it will also change what we mean by the word "work." I want you to realize that the changes that lie ahead are unavoidable. Society will become much more automated, the poor/middle-class as well as the super-rich will be shaken up, and pensions will never be what they once were. But it doesn't stop there. This revolution will have a direct psychological impact on our pride and our understanding of personal worth. In the current society, we value effort and hard work, but the future will appreciate other skills that are not yet discovered. The ability to adapt quickly, learn quickly, and be able to move where the opportunities are is what will separate the losers and the winner of this future era. Whatever your age is, it will be hard to move ahead because we will have to challenge the fundamentals that the schools of the past taught us. But we don't live in the past anymore. We need to be able to adapt to the future! So the question remains: are you learning skills that machines cannot do (yet)? Are you willing to work for the new jobs that do not exist? Are you going to be one of the survivors or the winners? Are you prepared to accept that, one day, you will not have the strength to live on your own?

Box 1: Is your job going to be automated?

Oxford University academics Michael Osborne and Carl Frey conducted a study on how susceptible each job is to automation. They based their analysis on nine fundamental skills: social perceptiveness, negotiation, persuasion, assisting and caring for others, originality, fine arts, finger dexterity, manual dexterity, and the need to work in a cramped workspace. In a very interactive website, BBC made the outcome of this study freely available for their readers. If you want to check the status of your current job (or the jobs of your friends and colleagues), browse the following website: http://www.bbc.com/news/technology-34066941. For the curious ones, this study suggests that the job of a telephone salesperson is the most likely to be automated (automation risk of 99%) and that of a hotel and accommodation manager or owner is the least likely to be automated (automation risk of 0.4%).[216][217]

Box 2: Do you know the pension plan that your best friends choose?

Personal finances are usually a private topic. We live in an era when everyone is connected to social media 24/7, but, at the same time, this increased transparency and sharing does not apply to private finances. It is bizarre to see a tweet about the pension fund that Mr. X chose for his retirement, and for good reason. Studies show that 57% of Americans consider their finances a private matter, and another 36% feel that finances can be discussed only with very close family and friends (typically, just one's spouse or partner). In other words, 82% of Americans say they know a lot about their partner's financial situation, while only 30% know about their children's, 13% their siblings', and a mere 8% their best friend's. This is quite a surprising fact if we take into consideration that the financial decisions we take will have an important effect on our lives. To be fair, there is a discernible shift among generations. Millennials are more transparent, and there's more openness to discussing money. However, discomfort is still the rule.[218] This reality has profound implications because people are assessed only by the financial institutions, which, at the same time, are the ones that sell them the products. An open debate is needed to increase transparency and empower competition (that will also benefit the customer).

Box 3: The mining sector in Spain

The European Union has issued a clear diktat: all the mines in Spain that are not profitable have to be closed in 2018 if they want to be helped by the government in this difficult moment for the sector. The EU has allowed the Spanish authorities to provide a 2 billion EUR fund to help those companies to dismantle. In

---

[216] (BBC, 2015)
[217] (Frey & Osborne, 2013)
[218] (Merrill Lynch, 2017)

other words, this funding has been released to close the mining business in the country. In the 1950s, there were more than 100,000 miners in Spain. Now, there are less than 3,000. The sector is simply going to disappear in the upcoming months. In economic terms,[219] these mines will never be able to compete with the mines from South Africa, Russia, the United States, or China that use cheaper labor and are fully automated. Globalization sometimes brings strange implications. The coal power plant of Compostilla (a small town in Spain) receives around 100 trucks every day from the harbor of Gijon. These trucks transport coal from all around the world. The cruel paradox is that there is a mine near the plant itself. This is one of the faces of globalization: thousands of jobs have been destroyed because prices are much cheaper in other countries of the world. This sector is definitely not the only one that is suffering from globalization. It is merely a new reality that we will have to face. If those miners and the cities where they live are not able to adapt to this new reality, they might have to face a deep and unpredictable social crisis in the region.[220]

---

[219] If we also added the environmental and social impact of this reality, the costs would definitely be higher.
[220] (El País, 2017)

ii. The role of governments and nations
To begin with

Globalization has widely helped the development of trade, commerce, and international business. However, this global city called Planet Earth has also brought some challenges at the national level. Countries and governments see their power and influence diminishing year after year. We don't live in the Middle Ages anymore, when trade between nations was rare and dangerous. International organizations such as the World Trade Organization (WTO) shape the new rules of the game. As you might expect, each country has a different degree of influence that depends on their share of the global trade market. Moreover, the exponential increase in the flow of goods, capital, and ideas is one of the most prominent economic trends in recent decades. A key driver of this phenomenon is cross-border production, investment, and innovation led by multinational corporations, which are some of the basic fundamental needs for this industrial revolution to succeed. Multinational affiliate sales as a share of world GDP have more than doubled in the past two decades.[221] They are competing with the countries themselves for influence on society. This balance of power leaves smaller nations behind, making it harder for them to set rules and make their voices heard. Will this new industrial revolution change the role of governments? And will the role of democracy be forever altered?

Influence of private companies on governments

There is a new paradigm in the world. Countries are being outperformed by multinational corporations that have a truly global power to influence decisions. Just look at the revenue ranking of the top 100 corporations and governments in Box 1. Clearly, big companies have the money to create waves in our society. But how will they make their influence felt?

---

[221] (Alfaro & Maggie, 2014)

Many corporations will try to fund lobbyists in order to defend their interests. On one side, this makes complete sense because the impact that those companies have on the lives of the people is sometimes more important than the impact of governmental policies. However, corporate lobbying brings with it a critical risk. Company lobbyists haven't been chosen by free vote, and they follow private interests that are sometimes very far from the public's interests. Companies will argue that lobbyists help to inform the government, and that the last word always belongs to the policymakers. And they are right. But it is sometimes too tempting to abuse your power and the generous budgets you have at your disposal, especially when your opponents don't have equal influential power. There are numerous examples that teach us that life is never black and white. Life is painted with a dark grayish palette.

One example that had a global impact was the distinct regulatory failures that led to the worst financial crisis in the United States since the Great Depression. The crisis played a significant role in the failure of key businesses, which led to declines in consumer wealth, estimated to be in trillions of US dollars, and a downturn in economic activity, leading to the Great Recession of 2008–2012 and contributing to the European sovereign-debt crisis. One of the key elements that started this crisis was the widespread availability of mortgages. Mortgages were given out with little review of recipients' qualifications, and the risks of default naturally increased. In the old times, banks made mortgages and held them. In the new era, banks and non-bank mortgage lenders sell their loans to other people. Investment banks packaged lots of mortgage loans into "Collateralized Debt Obligations" (CDOs) stamped with good ratings from agencies such as S&P, Fitch, and Moody's.[222] Those packages were then

---

[222] The job of these rating agencies is simply to give a rating (similar to a grade) to any financial product available on the market. The ratings are published by the credit rating agencies and are used by investment professionals to assess the quality of a financial asset. The better the rating is, the less risky the product is. The best grade to have is AAA. These rating agencies are funded by the banks themselves. This could explain suspicious ratings like the AAA of AIG or the AA of Lehman Brothers minutes

sold on Wall Street. As incredulous as it might sound, these operations went completely unregulated thanks to the power of lobbyists from the banking industry. Despite the supposed sophistication of the investors involved, no one took into account (or wanted to take into account) how *shady* the loans were or, more fundamentally, how high the certainty was that a large number of borrowers wouldn't be able to pay back their loans if the housing bubble popped.[223]

To be fair (and more optimistic too), private lobby groups have also brought great initiatives for the community. For example, the development of renewable energy is undoubtedly driven by the pressure of industry groups to develop those technologies. Let's take the case of the European Union. Before lobbyists came onto the scene, the vast majority of public subsidies went to fossil fuels. Now, the trend is moving toward subsidizing more renewable energy to make it attractive for investors. It is essential to mention that in the EU, energy consumption is still dominated by gas, coal, and nuclear energy. However, more financial support is being given to the renewable energy market (especially solar).[224] Clearly, this change in the mindset of politicians is not only because of the urgency of global warming but also because of the pressure coming from the renewable energy companies who know that this trend will bring business to their balance sheets.

Influence of intergovernmental organizations on governments[225]

This industrial revolution has made sure that governments now have new types of pressure coming from companies that are bigger than the governments themselves. However, in line with

---

before they collapsed, causing a global financial crisis of a magnitude never before seen in history.

[223] (The Huffington Post, 2011)

[224] (European Comission, 2014)

[225] The UN has used the term "intergovernmental organization" instead of "international organization" for clarity.

the new paradigm of the world, countries have gathered to form international groups to better deal with the challenges of the future global world. Organizations like the United Nations (UN), Organization for Economic Co-operation and Development (OECD), Organization for Security and Co-operation in Europe (OSCE), Council of Europe (COE), World Trade Organization (WTO), International Monetary Fund (IMF), International Criminal Court (ICC), G7, G20, and International Criminal Police Organization (Interpol) are some of the most popular examples. They have global visions built by their member governments. These organizations have been alive for many years, but their power and relevance are increasing with the development of global trade and the interdependencies among countries.

The bright side of these organizations is that they hold state authority, are issue specific, and allow multilateral cooperation. They are indeed the key for the development of countries that want to have global influence. The new industrial revolution will only increase their relevance and presence on the international stage.

But of course, international organizations come with a cost. In order to follow a common goal, trade-offs must be made. Having rules that are applicable worldwide means that the governments lose part of the sovereignty given to them by their people. Democracy and power of action might be affected. To mention some examples: seeking international funding for development might need some privatizations; tariffs might not be decided only by the heads of state; monetary instruments might be stopped by higher authorities, etc. States have to give up part of their sovereignty, which weakens the state's ability to assert its power. Moreover, membership is limited, and the organization usually prohibits the membership of private citizens. This makes intergovernmental organizations undemocratic by definition. Last but not least, inequality among state members creates biases and can cause powerful states to abuse their power. The power is moving from democratic governments to undemocratic umbrella organizations.[226] [227]

Resistance from the regional level

In opposition to the new emerging power of multinational corporations and intergovernmental organizations, it seems that a new protective pride has arisen. Thanks to technology, we live in a world with more and more connections and dependencies. In this new global city called Planet Earth, feeling unique and special as citizens of a country with a specific culture will become more and more difficult. As you might expect, some citizens have the feeling that decisions are being taken far from them and that the national sense of identity is losing its meaning. In some countries, those feelings of frustration are already well represented in political parties. The following are the most popular movements:

- In France, the Front National led by Marine Le Pen won almost 25% of the ballots in the European Parliament vote and holds 24 of France's 74 seats in the legislature. She ran for president during the 2017 elections and passed the first round. She lost against Emmanuel Macron in the second round, but the fact that a far-right representative made it to the last round of elections in a country known for human rights is a clear example of how societies are changing.
- In India, Narendra Modi, a member of the Hindu ultranationalist and anti-Muslim organization Rashtriya Swayamsevak Sangh (RSS) since his youth, was elected prime minister.
- In Japan, 15 of the 19 members of Shinzo Abe's Cabinet are members of Nippon Kaigi, a nationalist group that wants to turn back sexual equality, restore patriarchal values, and return Japan to a pre-war constitution.

---

[226] (Ankerl, 2000)
[227] (Stanford Encyclopedia of Philosophy, 2012)

- In the United Kingdom, the anti-EU UK Independence Party won the European Parliament election and saw its first two members elected to the House of Commons. Moreover, the UK decided by referendum to leave the EU.
- In the 2016 presidential election in Austria, the far-right Freedom Party's candidate Norbert Hofer won the first round, receiving 35.1% of the votes. However, he was narrowly defeated (49.7 to 50.3%) in the second round by the Green Party's candidate Alexander Van der Bellen.
- In December 2016, Donald Trump became the new US president promising, among other things, to build a wall between the United States and Mexico, ban Muslim immigrants (even if they were US citizens), void international treaties, and nullify multilateral agreements.

The relationship between the *renaissance* of nationalist parties and globalization is clear. Nationalist parties represent the alter ego of society's current direction. Nationalist supporters believe in a world less open to the outside, with a more protected economy and an enhanced sense of cultural unity and pride. They are the ones who will counterbalance and perhaps even stop this global economic expansion, shrinking the dreams of those who want the world to be a global connected village for their kids. Clearly, there are differences in the ways people envision the future, and those differences are feeding the resistance of those left behind.[228]

Final thoughts

The industrial and technological revolution will accelerate the development of a global market that is extraordinarily connected and interdependent. But of course, it will come at the cost of shifting sovereignty from the countries toward companies and intergovernmental organizations, feeding resistance from certain segments of the population. This revolution will change the expectations that citizens have of their democratically elected

---

[228] (Bershidsky, 2014)

government and disrupt the definition of citizenship. New policies are increasingly being decided in rooms that are more and more cut off from the citizens. And this distance is expected to increase with the development of the new industrial revolution. Today, we have more speakers that we've ever had before to fight against what we believe is unfair, but ironically, our voice is diluted as we become smaller and more insignificant in the decision-making process. Do you want to trade sovereignty for safety and development? Do you accept and integrate people and investments coming from other lands? What place should your country have in the global world?

Box 1: When a company is stronger than a government

When we read the profit and loss statements of the most important multinationals in the world, it is difficult to put those values into perspective and see the big picture. The NGO Global Justice Now has helped us by comparing government revenues (from the CIA World Factbook) with corporate turnover[229] (from Fortune Global 500). The results are astonishing and a little bit scary; among the world's top 100 institutions in 2015 were 31 countries and 69 corporations. The list is below in Table 06. Food for thought. [230]

RANK	COUNTRY /INSTITUTION	USD BN[231]	RANK	COUNTRY / INSTITUTION	USD BN
1	United States	3251	51	Hon Hai Precision	141
2	China	2426	52	General Electric	140
3	Germany	1515	53	China State Construction	140

---

[229] The corporate turnover is the company's total revenue, from the invoices, cash payments, and other revenues. Corporate turnover represents the value of goods and services provided to customers during a specified period - usually one year.
[230] (World Bank, 2016)
[231] Government revenues and corporate turnover in billion USD.

4	Japan	1439	54	Amerisource Bergen	136
5	France	1253	55	Agricultural Bank of China	133
6	UK	1101	56	Verizon	132
7	Italy	876	57	Finland	131
8	Brazil	631	58	Chevron	131
9	Canada	585	59	E.ON	129
10	Walmart	482	60	AXA	129
11	Spain	474	61	Indonesia	123
12	Australia	426	62	Allianz	123
13	Netherlands	337	63	Bank of China	122
14	State Grid	330	64	Honda Motor	122
15	China Petroleum	299	65	Japan Post Holdings	119
16	Sinopec	294	66	Costco	116
17	South Korea	291	67	BNP Paribas	112
18	Shell	272	68	Fannie Mae	110
19	Mexico	260	69	Ping an insurance	110
20	Sweden	251	70	UAE	110
21	ExxonMobil	246	71	Kroger	110
22	Volkswagen	237	72	Société Générale	108
23	Toyota Motor	237	73	Amazon.com	107
24	India	236	74	China Mobile	107
25	Apple	234	75	SAIC Motor	107
26	Belgium	227	76	Walgreens Boots	103
27	BP	226	77	HP	103
28	Switzerland	222	78	Assicurazioni	103
29	Norway	220	79	Cardinal Health	103
30	Russia	216	80	BMW	102
31	Berkshire Hathaway	211	81	Express Scripts	102
32	Venezuela	203	82	Nissan Motor	102

33	Saudi Arabia	193	83	China Life Insurance	101
34	McKesson	192	84	J.P. Morgan Chase	101
35	Austria	189	85	Gazprom	99
36	Samsung	177	86	China Railway	99
37	Turkey	175	87	Petrobras	97
38	Glencore	170	88	Trafigura Group	97
39	Industrial Bank of China	167	89	Nippon Telegraph & Telephone	96
40	Daimler	166	90	Boeing	96
41	Denmark	162	91	China Railway Construction	96
42	United Health	157	92	Microsoft	94
43	CVS Health	153	93	Bank of America	93
44	EXOR Group	153	94	ENI	93
45	General Motors	152	95	Nestlé	92
46	Ford Motor	150	96	Wells Fargo	90
47	China Const. Bank	148	97	Portugal	90
48	AT&T	147	98	HSBC Holdings	89
49	Total	143	99	Home Depot	89
50	Argentina	143	100	Citigroup	88

TABLE 06: TOP 100 ECONOMIES IN THE WORLD IN 2015 (INCLUDING GOVERNMENTS AND CORPORATIONS)

Box 2: Common Agricultural Policy in the EU and free trade

The Common Agricultural Policy, or CAP, is a system of subsidies paid to EU farmers. It has two main purposes:
Guarantee minimum levels of production in the EU

Ensure a fair standard of living for those working in the agriculture industry

In 2013, the budget for CAP was an outstanding 43% of the European Union's total budget: €57.5 billion out of €132.8 billion. Most of the CAP budget is paid directly to farmers. Agriculture in the EU generates 1.6% of EU GDP and employs only 5% of EU citizens. Setting aside the fair/unfair distribution of those subsidies and the cost of this policy, the EU is facing criticism from the international community. The EU cannot use all its agricultural products, so it sells them cheaply to the Third World. This undercuts local farmers, who cannot compete with the heavily-subsidized imports.[232] [233] This is one clear example of:

How impressive the power of the lobbyists sitting in Brussels who work for the giant European agri-businesses is.
How countries gather together in groups (like the European Union) to be stronger against outside competition.
How even if the developed countries give lessons to the world about free trade and free markets, they fake prices[234] to compete globally against developing countries.
How protectionism still has a long way to go in our capitalistic societies, even if we move toward global trading.

Box 3: IMF to Developing Nations: Privatize!

The International Monetary Fund[235] is one of those organizations funded by governments that is working internationally to pursue the development of global wealth. Even if, in theory, IMF is very aligned with the values of globalization and common prosperity,

---

[232] (BBC, 2013)
[233] (Jeffery, 2003)
[234] By subsidizing activities, the government is virtually changing the final price of a good.
[235] Also known as IMF, this fund was conceived at a UN conference in Bretton Woods in 1944 to build a framework to avoid new Great Depressions like the one in the 1930s. The IMF's primary purpose is to ensure the stability of the international monetary system. A core responsibility of the IMF is to provide loans to member countries experiencing actual or potential balance-of-payment problems. Unlike development banks, the IMF does not lend for specific projects.

this fund has received public criticism. One of the main objectors is Joseph E. Stiglitz.[236] According to him, the IMF interventions in emerging countries follow a free-market formula copied from formulas that would be effective in the United States. The IMF provides a quick "shock therapy" to market economies without first establishing institutions to protect the public and local commerce. When privatizations are promoted without land reform or strong competitive policies (like the United States does), countries can fall under the management of large businesses run by organized crime and a neo-feudal society without a middle class. Without strong progress at the forefront of the policy, the help could very well be counterproductive. The consequences will be escalated levels of debt, weakened policy credibility, and greater difficulty adjusting in the future.

In the words of Stiglitz, the IMF also foisted premature capital market liberalization without institutional regulation. The consequence was a destabilization of entire developing economies by causing massive inflows of "hot" short-term investment capital. Moreover, when inflation rose, the IMF's loan conditions imposed fiscal austerity and dramatically rising interest rates. This led to widespread bankruptcies without legal protection, massive unemployment without a social safety net, and the prompt withdrawal of foreign capital. With loans defaulted and entire nations thrown into economic and social chaos, the IMF rushed bailouts directed mainly to foreign creditors. As a result, citizens of developing nations carried much of the costs and few of the benefits of IMF loans, and a moral hazard ensued among the financial community. Foreign creditors made bad loans, knowing that if the debtors defaulted, the IMF would pick up the tab. Meanwhile, the IMF urged cash-strapped countries to further privatize—in effect selling their assets at a fraction of their value to raise cash. Foreign corporations then

---

[236] Joseph E. Stiglitz is the recipient of the Nobel Prize in Economic Sciences (2001), a former chief economist of the World Bank, and former member and chairman of the US president's Council of Economic Advisers. He is known for his critical view of the management of globalization and of international institutions such as the International Monetary Fund and the World Bank.

bought up the assets at rock-bottom prices. Putting aside the potential political influence that may appear in these types of loan approvals, the key issue to be raised is this: how much right should non-democratic organizations like the IMF have to interfere in the economy of a sovereign country? This example clearly shows the trade-offs coming from the development of a globally connected and interdependent world.[237] [238]

---

[237] (Stiglitz, 2002)
[238] (International Monetary Fund, 2017)

iii. IT and your company

To begin with

If you read the annual report of any company in any sector, it will emphasize innovation and plans to outperform competitors in managing the new era. The new industrial revolution has raised the standard of what a company should be, especially in terms of IT. IT is not a single organizational unit anymore. It has become a support unit that has to be everywhere all the time, like the oil that lubricates the whole company. The truth that nobody seems to admit publicly is that IT integration is complex, takes time, is unpredictable, requires a culture friendly to change, and represents a significant investment. But of course, that's why IT is powerful. It has the potential to transform the company's success. Let's take a look at the main barriers that companies are facing and how they are contributing to this fast industrial revolution.

Different organizational units now have to talk... Isn't that scary?

As crazy as it might sound, different departments will now need to talk: "cross-functional communication" as some people call it. Many companies struggle to break the cliques formed by their departments. The reality is that there are walls between different functions that make it difficult to communicate and establish a common digital strategy for years to come. The digital revolution will not only affect one specific department but the organization as a whole. This push for radical transformation and the willingness to change is simply absent or very low in many companies. But if they don't change their business models and adapt to the new digital reality, they will disappear (even if they are a Fortune 500).

We are facing challenges that require strong and innovative leadership as well as a flexible workforce, both at the same time. This industrial revolution is not only based on interdependencies between countries, but is also, in fact, based on interdependencies between the different departments of your company. And the glue holding those dependencies in place is the IT department, which will morph from "IT department" to "key support for every department." Data merged with technology will be king. Like it or not, we will have to be good at new technologies if we want to keep our jobs. We will not be able to say "technology is not for me" or "I am too old to learn all those skills."

In the development of IT, a corporate culture of continuous training will play a decisive role. The management will have to be brave enough to move toward radical transformation and new ways of working. The reality is that the leaders who pull the strings tend to lack the courage to make changes or face resistance from their employees. Change is always scary. But if change is combined with a lack of a clear business case to justify the investment and with the uncertainty of the current times, the combination becomes too risky. Changes are withdrawn from the board's agenda, and the board prays that no tech company will enter their industry. Because let's be clear. If they do, it will be too late. Don't believe me? Ask Polaroid Corporation. They were a pioneer in digital imaging in the 1960s. However, they didn't make the necessary investments to hold that lead in the 1990s, when digital photography overtook film. The company went from peak revenue of $3 billion in 1991 to declaring bankruptcy in 2001 when its brand and assets were sold off. The "new" Polaroid formed as a result declared bankruptcy again in 2008, resulting in a further sale and the present-day Polaroid Corporation.[239] Do you still believe that digital is not for you?

Lack of talent

Many business leaders argue that the talent required for this digital transformation is scarce and therefore expensive. The lack of data scientists, for example, is a clear risk for many businesses that want to exploit the potential of big data. Some years ago, few people understood the value of a skill set that combines computer science, statistics, engineering, business insights, and strategy. But in the current times, it is a key requirement. Indeed,

---

[239] (Strategy+business, 2017)

data science has given consumers personalized shopping recommendations on eBay and Amazon, customized entertainment recommendations on Netflix and Spotify, and tailored content and advertisements on Facebook, Twitter, and Pinterest. Almost all the successful online businesses that we know of need the application of data analytics. Just at a glance, data production is expected to exceed 240 exabytes[240] daily by 2020. The companies that figure out how to turn those bytes into insights will become the Googles and Facebooks of the future. The problem is that there's a global talent shortage, and the demand for specific digital talent continues to grow rapidly, far outpacing the anemic growth in supply. If we follow current job trends, by 2018 the number of data science jobs in the United States alone will exceed 490,000, but there will be fewer than 200,000 data scientists available to fill these positions. Globally, demand for data scientists is projected to exceed supply by more than 50% by 2018. Data science is one of the most notable examples of a rare skill, but in general the digital skills shortage will become a main issue in the years to come. We will have data with huge business potential, and we will not be able to use it due to the lack of talent.[241] [242]

Data ownership and cybersecurity

Companies don't have a choice; they have to go digital. But they face a new dilemma: should they go for a third-party IT provider, or should they develop the tools internally?

In old times, knowledge was stored in the brains of engineers and the books of the company library. Today, companies have to outsource part of their processes to IT suppliers if they want to remain competitive. But there's a catch. On the one hand, external companies with unknown interests could potentially have access to sensitive data that could be very valuable for your

---

[240] One Exabyte is a 1-followed-by-18-zeros bytes. Or, to put this number in perspective, the amount of data that can be stored in 250 million DVDs.
[241] (TechCrunch, 2015)
[242] (McKinsey & Company, 2016)

competitors. Moreover, it is very likely that your competitor is using the same IT supplier that you are using, in one way or another, leading to potential leaks that can be critical for your interests. On the other hand, corporations don't have a choice if they don't want to explode their budgets. They need those IT suppliers to remain alive in the market. Data ownership will be one of the most critical decisions that company managers will have to make to stay relevant. The world is becoming interconnected and, therefore, totally interdependent.

But fear does not only come from your partners. It also comes from outsiders who see an opportunity to make some money out of your connected company. Nowadays, an office cannot function without an internet connection, but that connection is an open door to the outside world. Cybersecurity will, without any doubt, become one of the most important industries in the near future. Multinationals are implementing worldwide systems that *talk* among themselves in order to use technology to its fullest potential. The security required for those *dialogues* is a critical investment that companies will have to make. But the potential harm of cyber-attacks is far more threatening than some industrial espionage. Let's not forget that the Internet of Things (an industry that will have an annual economic impact of $3.9 trillion to $11.1 trillion worldwide by 2025) will make things talk among themselves. All those conversations will move in a wireless environment where messages can be intercepted. The problem with security in general and cybersecurity in particular is that customers and end-users tend to view it as a commodity and, therefore, are not willing to pay a premium for it. It is usually seen as a cost before a need, even if the potential damages are far from cheap. Cybercrimes will cost the world an estimated $6 trillion annually by 2021, up from $3 trillion in 2015. At the same time, companies and humans will become even bigger data producers than they are now, feeding a global interconnected world. The extent to which we can trust the digital platforms will define our digital fate. Never underestimate the impact of cybersecurity.[243] [244] [245]

Final thoughts

In other words, everyone seems to agree that the future of work moves toward more IT. But the reality is that, in many cases, the implementation phase is lacking a strong vision from management, willingness to cooperate from the workers, and the talent to configure IT systems. Besides, critical threats to the company's survival come along with the word "digital," making change even more difficult. The digital era will bring us innovation and progress, but it will also bring challenges that we are not yet ready for. The ability of corporations to tackle these obstacles will define the winners and the losers of this new era.

Box 1: The Kaggle acquisition

Kaggle is an online service that hosts data science and machine learning competitions that draw hundreds of thousands of data scientists. Google acquired it in 2017. The objective is clear; Kaggle will give the tech giant the ability to reach out to the AI and data science community directly. Buying Kaggle will also help Google's recruiting team. Google needs to ensure they can attract top-notch talent if they want to stay meaningful in the market. This acquisition is just another example that confirms that data scientists are becoming a scarce and precious resource and that artificial intelligence is seen as the future of the industry. In other words, Google just bought a pool of potential brains.[246]

Box 2: eBay me!

One of the largest online security breaches in history left the passwords, names, addresses, telephone numbers, and financial information of 233 million eBay users worldwide in the hands of

---

[243] (McKinsey & Company, 2017)
[244] (World Economic Forum, 2011)
[245] (Investing News, 2018)
[246] (TechCrunch, 2017)

hackers. Security experts warned that the two weeks eBay took to inform their customers of the breach increased the risk of criminals using the information to create fraudulent bank accounts and/or supplying the identity of the users for illegal practices. This is just an example of the power and impact that hackers can have worldwide, even when companies encrypt their data. If global market leaders such as eBay are not able to secure themselves against hackers, how much can we trust all the other IT businesses with our private data?[247]

Box 3: Tape against espionage?

In 2016, a trendy picture surfaced that represents how serious privacy and cybersecurity are. It was a picture of the CEO of Facebook: Mark Zuckerberg. The founder of the social media platform made a post on his profile announcing that Instagram had reached its latest milestone of half a billion monthly active users. The curious detail that was hard to miss was that the camera and the microphone of his computer were covered with tape. This may sound paranoid, but if the CEO of a multibillion-dollar IT company doesn't trust the security of his computer, maybe we should think again about cybersecurity. [248] [249]

---

[247] (The Telegraph, 2014)
[248] Unfortunately, I couldn't add this picture in this book because of copyright infringements. However, it can be easily found on the internet.
[249] (The Guardian, 2016)

Technology will completely shape the job market and the economy. Developing the workforce skill set to align with technology and change will be the new norm. Governments will be disrupted by the structures they created and their own multinationals, and will have to find their place in this new reality. The future will be bright for those who can adapt quickly. Adaptation will be the critical skill to prevent yourself from being left behind. Are you ready? Is your family willing to take the adaptation path? Is your circle of influence qualified to make a change in the future?

B. How will **you** be afraid of your society?

Indeed, it is easy to be scared of the new reality that we are facing. And the future doesn't offer any easy solutions. Digital disruption will have an impact beyond companies. Like it or not, it will directly affect our societies too. We tend to forget that the convenience of technology has a cost. We will become afraid of the comfort that we have created. Society will be more connected and therefore more out of your control. Technology will even affect our definition of freedom and free will. Not convinced? Let's take a look.

i. Privacy as a core risk

To begin with

Privacy is by far one of the most important risks that technology is bringing into the picture. We tend to forget it, but it is a major challenge. Any information that we send through the internet is stored somewhere, encrypted somehow, and guarded by somebody. The problem is that the somewhere, somehow, and somebody are usually unknown and therefore represent a threat to the information itself and our privacy in general. And of course, this information might also be sold to third parties. With the internet, the mobile phone revolution, and the upcoming smart house expansion, privacy and security will become more and more important. What I am afraid of is that the needed social debate to address this vital issue is not yet on any political agenda. Always remember that the information that you put online is difficult to erase and that it could potentially be used against you in the future.

The new advertising industry and the "I have nothing to hide" argument

We are giving away data every single day. This data is automatically stored in the so-called "cloud." The cloud is simply a bunch of servers that store data and, in some cases, try to transform those 1s and 0s into valuable insights that can then be used for targeted ads. Companies and governments want to know everything about you: age, location, habits, number of friends, favorite artists, sexual orientation, marital status, your last purchase on Amazon, your holiday destination... Simply because this information has a value—whether it is to sell you a new Nissan or to know which party you will vote for in the next elections.

Internet advertising revenues represent a huge market. In the United States alone, revenues increased by 22% from 2015 to 2016 to reach the record of $72.5 billion, surpassing for the first time in history the $69 billion spent on TV ads.[250] Advertising is key and represents more than three-fourths of the total revenue of the main tech players. Your data is extremely lucrative.

When I explain those numbers in conferences, I always get the same reaction, something like: "Yes, I know that the Googles and Facebooks of the internet are taking a look into my data to give me better ads, but I love it! I have a much better advertising experience, and it is extremely convenient." And here is where it becomes interesting. It may feel amazing to receive an ad notifying you that your favorite DJ is coming to your city. But wait, is it?

The problem is the price you paid for this advertisement. In the example, you provided your location and your personal music taste to a major tech corporation. You shared information with a global profit-making company that you probably wouldn't tell someone you met at a party. And you don't stop there. You also shared your public and private pictures from your last holiday, your "I miss you" texts to your now ex-girlfriend, your 3 a.m. calls, your angry comment about your boss, your passwords, your spending behavior, your Hello Kitty purchase on Amazon... This type of information is extremely valuable for advertisers but could be used against your will by somebody that would like to hurt you or take advantage of you. You never know how the future will look and what exactly companies will do with your data.

---

[250] (Business Insider, 2017)

The statement from the audience continues: "But anyway, I am not a terrorist or a murderer. I have nothing to hide." Here is where people get more nervous: "I have nothing to hide." Let me tell you something: yes, you do have things to hide. Thank God you do! This doesn't mean that you are a criminal; this simply means that there are parts of your life that you want to keep for yourself. Let's try a social experiment. Because you have nothing to hide, could you please give me your email and its password so I can take a look into it? No worries, I will not send or delete anything. I simply want to see what you write, to whom, at what time, from where, and what their replies are. You would probably refuse to give me this information even if I was your best friend. Why? You have nothing to hide! The answer is simple. There is information in your email account that you don't want to share with anyone. This is the basis of privacy and of freedom too. Edward Snowden described this dilemma with the following quote: "Arguing that you don't care about the right to privacy because you have nothing to hide is no different than saying you don't care about free speech because you have nothing to say. When you say, 'I have nothing to hide,' you're saying, 'I don't care about this right.'"[251] This new era will challenge the right to privacy in our new share-it-all society.[252]

Government surveillance programs and democracy

Surveillance programs have always existed. From Carnivore and ECHELON to PRISM, from Boundless Informant to FISA orders, secret programs control and manipulate the societies of the country. They are not only meant for terrorists; they are used to track everyone. However, privacy is a right granted to individuals that secures the freedom of expression, association and assembly, all of which are essential for our democratic society. A loss of privacy leads inevitably to a loss of freedom.

First, we have to understand the power of aggregation. Bits of information from different sources could potentially be

---

[251] (TechCrunch, 2014)
[252] (TED, 2014)

aggregated into a profile that could give a clear image about any individual. Your interactions with the world around you can reveal your political and religious beliefs, your desires, sympathies, and convictions. This *virtual you* could even predict your behavior with enough information and the use of artificial intelligence and big data analytics. The government understands this power. That's why they track the information held in the servers of the tech companies that we all use and love. This surveillance poses two main threats to democracy:

-Your <u>freedom of expression</u> is threatened by the surveillance of your internet usage. The knowledge that you could be under surveillance can make you less likely to research a particular topic, and your thoughts can be pushed in one direction as a result. In other words, you self-censor, causing the rest of us lose your perspective, and the development of further ideas is stifled.

-Your <u>freedom of association</u> is threatened by the surveillance of your communications, and your freedom of assembly too, by the tracking of your location through your smartphone. The social change brought by activists and campaigners could be at risk.

Governments use this information to potentially control their citizens. They are spying on their enemies, allies, and their own population. Just ask the German chancellor Angela Merkel how the NSA tapped her phone calls.[253] The convenience technology brings its users is also extremely convenient for surveillance programs. And this trend will continue to grow. If we believe in democracy, we need to fight for our right to privacy.[254]

Criminal offenses

With the rapid diffusion of technology, the number of crimes has increased dramatically. The problem is that privacy itself (or the lack of it) can become a tool for criminals. There are thousands of cases in every country. The shocking fact is that the criminals are

---

[253] (The Guardian, 2015)
[254] (Medium, 2016)

not professional hackers who breached your account. They are people with average social media knowledge who can check your profile and know your favorite bar, snag your drunk picture from last night, know the restaurant you will go to next Saturday, and the name of your partner, among a lot of other private information. This information can easily be used to understand the victims and their behavior and to ultimately commit the crime. All sorts of criminal offenses are possible and the convenience of technology is also making it easier to commit a crime. Here's a brief list:

Scams: The strategy is to convince an individual to click on a link that would interest almost anyone (e.g., a cat video) or that would offer you a fantastic prize (e.g., free iPhone!). Clicking the link will then install a backdoor to your computer so hackers can monitor your activity and take control of your laptop.

Cyberbullying: The concept is simple: bullying somebody though a social media platform. Unfortunately, it is a common practice among teenagers. Several teens have been driven to suicide by relentless cyberbullying.

Stalking: Even if it is usually considered a joke (stop looking at your ex-girlfriend's profile!), stalking typically involves harassment through messages, written threats, and other persistent online behavior that endangers a person's safety.

Harassment: It's not uncommon for sex offenders and sexual predators to prey on unsuspecting victims on Facebook or even create a fake profile to pose as a college student.

And many more...[255]

Here is where your responsibility takes on an important role. We are giving away private information for free. It might be used against us. As innocent as it might look, we tend to forget that

---

[255] (Security Affairs, 2012)

privacy is one of the critical elements of our social life. Digital education is key. We must ensure that we treat social media platforms carefully to protect ourselves and our loved ones. If a problem arises, it is imperative to speak up quickly (see Box 1).

Final thoughts

Threats to privacy will be one of the main challenges to overcome in the following years. It is important to realize that this new always-connected era will not only change our definition of personal digital space but will also change our perception of freedom and our right to have secrets. We are creating platforms that have incredible potential, but we are giving away our privacy. The future of privacy will depend on how we choose to use the technology at our disposal.

Box 1: What can be done about criminal cyber offenses?

I want to be extremely clear about this point. It is crucial that as a society we take action and protect ourselves against the problems that could be created by digital societies. And this part is not only for you but also for your circle. It is your responsibility to protect yourself and help your peers if you see any kind of criminal action being perpetrated against them.

If you find yourself or somebody else in a situation of abuse, the first thing you must do is speak to the local authorities. It is important to mention that cyber offenses are criminal acts and are punishable by law. Next, the social media platforms usually have their own weapons to fight against abusive behaviors. Don't hesitate to use them. You have the right to keep your privacy safe. Afterwards, it's important to educate yourself. There is a lot of information online about the legal actions to take if you are targeted by cyber criminals and the rights that you have as a citizen. One particularly good website is *smartsafe.org.au*. Usually your government also has hotlines and information available. Last but not least, it is important to have social support. Don't hesitate to share the problems you encountered with your family and/or trusted friends. If you don't feel comfortable sharing your experience in person or cannot find anybody to turn to for help, don't hesitate to contact national hotlines, reporting websites,[256] or even me. Moreover, if you have a friend who is going through a difficult situation because of cyber harassment, be sure you help him or her without judgment. You never know when it might happen to you.

Box 2: Revenge pornography

Revenge pornography is the distribution of sexually explicit images or videos of individuals without their consent. This material may be recorded or filmed with or without the knowledge and consent of the subject. Its possession may be used for different purposes, from blackmailing to coercing the victim. The material is usually accompanied by information about the victim to identify him or her. The objective is frequently to humiliate and intimidate. This practice has unfortunately become more popular in the last few years. According to a study published by the Data & Society Research Institute, 1 out of every 25 Americans (around 10 million people) are either threatened with or are victims of non-consensual image sharing. It is clear that an open dialogue is needed regarding digitalizing intimate moments. The consequences might be dramatic… Just for your information, it is worth knowing that in June 2015, Google announced it would remove links to revenge porn on request. Microsoft did the same one month later. Together, both companies account for almost 90% of the internet search market, and they both have placed forms online for the victims (even if usually it is too late). [257]

Box 3: Google Search is the only person that you never lie to

You feel completely comfortable sharing your most profound questions and desires with your search engine, but you may not realize that this information is being tracked and stored on servers around the world. We kind of trust a privately owned server and algorithm more than the advice of a good friend or

---

[256] Just google "hotline cybercrime <name of your country>" to find out more.
[257] (Data&Society Research Institute, 2016)

partner. Maybe this is something to think about. Just as an example, if you are a Google user, go to the following link: https://myactivity.google.com/myactivity. You will be surprised...

ii. Different speeds in tech and those left behind

To begin with

Several indicators (GDP, HDI, inflation, etc.) are widely used to compare countries' performances. A country with high GDP per person is usually aligned with a high technological development, but it might be the case that new indexes will have to come into the picture to better predict our technological adaption in the future. One interesting concept would be the ICT Development Index or IDI (check Box 1 to know more). The IDI monitors specific IT indicators to define the development of a country in the technological sense. This type of index will help us understand what the word "development" means for a nation. We have discussed how technology is and will become the key factor of differentiation for countries, companies, and individuals. However, even if I believe it is a clear enabler, those different speeds of acquisition will also define the people left behind and the losers of this new era.

The war on intellectual property: assets becoming digital

Technological know-how will become the main competitive advantage of countries. Having better access to the latest software (including artificial intelligence) will be the one key element that will differentiate the world's economies. However, software is something intangible and easy to copy. Let's look at an example. Imagine an oil company that wants to open a refinery in the country X. The company will have to send their engineers, machines, construction materials, etc., in order to build the factory and extract the oil. Moreover, they will need to have agreements with the local government, access to electricity, access to roads, etc. However, imagine now the same company using the latest possible technologies. They could send the whole factory previously made by robots to the country itself with the help of drones or helicopters. They could even print it there! And, because this factory is almost completely managed by artificial intelligence and automation, it could start to operate immediately. This modus operandi will decrease costs and increase speed of action. Isn't that great?

But there is a problem. Code is a pretty intangible asset.[258] And, by definition, intangible assets are difficult to control and legislate–especially if you take into consideration that artificial intelligence will be programming the majority of the code. As strange as it may sound, programmers will also be partially automated by artificial intelligence that will write parts of their code.[259] In the current times, intellectual property protects companies and individuals against fraud. It is definitely not a perfect system, but it protects inventors and developers. However, in the upcoming years, intellectual property will play a crucial yet complex role, especially with software that is made by machines themselves. As usually happens, legislation will be slow in tackling these challenges, which will inevitably leave some countries behind. The ones that fail to bring legal protection in the digital field to foreign companies will not be able to shine. Because globalization and this industrial revolution go hand in hand, a global coordination is needed to conveniently protect software innovation and global innovation in general. The countries that do not have strong legislation will be on the losing side.

Acquisition speeds

Indeed, we will need technology to survive. Speed of software acquisition will become one of the key factors of differentiation among companies and sectors. However, it is difficult to change people's mindset. The older generations are not used to new ways of working. The rules of the game are changing, and changing fast. Barriers to enter industries are decreasing and disruption is the new rule. Business is now done through WhatsApp, opportunities for remote work are expected, documents are sent electronically through Google Drive, hierarchy is less strict, and you can check out the résumé of the person you are talking to through LinkedIn. Companies that don't

---

[258] Under most circumstances, computer software is classified as an intangible asset because of its non-physical nature. However, accounting rules state that there are certain exceptions that permit the classification of computer software as Property, Plant, and Equipment (PPE).
[259] (Futurity, 2018)

follow these trends will no longer be competitive. Companies are trying (with more or less success) to bring these new skills to their employees. The problem is that, when those companies are not competitive anymore, they will run out of business. Their employees will have to find a new job, but they will simply not have the skills (nor the suitable age) needed to find one. They will have the theoretical knowledge of the industry, but not the new technology-focused culture needed to succeed in it. In an extreme example, how does having 30 years of experience in the industry help you to understand a technology that is six months old in a youth-oriented culture? Of course, some will argue that experience is everything. And they are right. But think again about the nature of jobs in the future:

- 47% of the current US jobs are at risk of disappearing in the next 25 years according to an Oxford study.[260] The trend is similar in other developed countries.

- By 2030, 75 million to 375 million workers (3 to 14% of the global workforce) will need to switch occupational categories. Moreover, all workers will need to adapt, as their occupations evolve alongside increasingly capable machines.[261]

Besides, specific industries and positions will disappear because of automation and artificial intelligence (e.g., auditing, transportation, translators, secretaries…). It isn't necessarily bad news because there will be new jobs and industries created. But the problem remains the same. If society is not able to learn the new skills required, companies will just be left behind bringing along systemic unemployment. If the workforce is not trained in time to face the new reality knocking at our doors, it might be too late to make amends.

---

[260] (The Guardian, 2017)
[261] (McKinsey & Company, 2017)

Technological illiterates

Last but not least, a new concept will be introduced to our day-to-day vocabulary: technological illiterates. In the current times, reading and writing are defined as the most basic skills that our educational system gives to its students. Literacy has been traditionally defined as the ability to read and write. Depending on the source you refer, there are different definitions for literacy and the level of understanding required to consider an individual literate. The global literacy rate for all people aged 15 and above is 86.3%; for men, the figure stands at 90.0%, and for women at 82.7%. However, this rate depends a lot on the country. For developed nations, this rate is usually more than 99%, but it plummets to 64% when moving to sub-Saharan Africa. Over 75% of the world's 781 million illiterate adults are found in South Asia, West Asia, and sub-Saharan Africa, and women represent almost two-thirds of all illiterate adults globally.[262] [263] Even if the number of illiterate people is decreasing, there is a clear correlation between the level of economic development in a country and its literacy rate.

But the needs of society are changing. In the following years, it will be more important to know how to manage an iPad than how to conjugate irregular verbs. It is currently more important to own a smartphone than to own a library full of books. We are witnessing more and more services go fully digital and the trend is increasing exponentially. Just to give an example, banks are closing branches and replacing their services with apps and artificial intelligence. Will this mean that branches will disappear? Probably not. However, financial opportunities for technological illiterates will be decreased and, at some point, the technologically illiterate people will be left behind with little chance to succeed. At this point, societies will face the most dramatic side of technology with a dilemma that has a complicated answer. There is no doubt that technology is an enabler of humankind's development, but those who do not

---

[262] (UNESCO, 2015)
[263] (The World Factbook, 2015)

follow will be left behind. Does this ring a bell? In terms of job destruction, the risk is not that people will be unemployed; the risk is that they will become unemployable. [264]

Final thoughts

This industrial revolution may leave people, companies, and even countries behind, but it will also change our understanding of the word "development." It will change what we consider essential in education and corporate training. Even more, a need for global legislation and multilateral agreements will pop up. Changing the mindset of lawmakers and regulators will be key in succeeding and taking advantage of this new era. Is your legislation ready? Is your company ready, or does disruption not happen in your industry? Are you ready to embrace the new era, or is technology not for you?

Box 1: IDI or ICT Development Index

The ICT Development Index is an index published by the United Nations International Telecommunication Union. It is based on 11 ICT (Information Communication Technologies) indicators, grouped into three clusters:

ACCESS:
- fixed-telephone subscriptions per 100 inhabitants
- mobile-cellular telephone subscriptions per 100 inhabitants
- international internet bandwidth (bits/s) per user
- percentage of households with a computer or similar device
- percentage of households with internet access

USE:
- percentage of individuals using the internet
- fixed (wired) broadband subscriptions per 100 inhabitants
- wireless broadband subscriptions per 100 inhabitants

---

[264] (Wired, 2017)

SKILLS:

- adult literacy rate (the percentage of the population that is above 15 years of age and can read and write simple statements and do simple arithmetic)
- gross enrollment ratio, secondary level (percentage of eligible individuals who are enrolled in a specific level of education)
- gross enrollment ratio, tertiary level (percentage of eligible individuals who are enrolled in a specific level of education)

In the near future, we might hear this index more often in the news because it will become key to understanding how technologically ready we are.[265]

Box 2: Improving labor market dynamism

Workers in countries with more fluid labor markets find work more quickly and obtain jobs that are a better fit for them. However, this expected dynamism is facing a different reality. In advanced economies, there is evidence that labor markets are becoming less dynamic, with fewer people switching jobs and with geographic mobility decreasing. Digital solutions can remedy these problems. For example, digital platforms will make labor markets more transparent and improve job matching or flexible work options that may enhance the re-employment of displaced workers. However, the mindset has to change. More digitalization requires more flexibility and more adaptability to emerging technologies. Somehow, the workforce must adapt to the changing environment. There will be very few companies that possess this changing skill in-house; therefore, this flexibility will need to come from the workforce itself. Get ready to be dynamic![266]

---

[265] (International Telecommunication Union, 2017)
[266] (McKinsey & Company, 2017)

Box 3: Mirmiki, teaching our elders how to use new technologies

Mirmiki is a promising Spanish start-up that envisions helping our elders (and people who are not that old!) use the new technologies on the market. It offers private classes on handling mainstream platforms like WhatsApp and Gmail. Those applications might seem simple, but for people past a certain age, technology is not as easy as it is for us. This start-up is an example of what society currently needs. More and more people are being left behind by technology, and it is the duty of each one of us to prevent this from happening.[267]

---

[267] (Mirmiki, 2018)

iii. Ethical issues

To begin with

Probably, one of the first pictures that comes to our minds when we think about technology and ethics is an army of robots invading the world because we have failed to set the correct boundaries. It is necessary to differentiate between the potential of technology and our ethical limits. In the current times, ethics usually come too late because politicians try to dodge decisions about controversial topics.

Each new technology that appears requires a period of reflection and social debate to set the legal boundaries and adapt the current legislation to the new reality. For example, autonomous vehicles have been functional for years (way earlier than the Google car). However, governments are still thinking about how to set the correct legal framework for autonomous cars. Questions, such as who is responsible in case of an accident, need to be regulated to establish the baseline for the development of this technology. Of course, other delicate questions, like the place of the transportation industry in the future, come along too. Ethics is a much broader topic than the boundaries of the technology itself.

The future of life as a source of controversy

By far, one of the most controversial ethical topics is in the field of medicine. We've discussed the potential of medicine (especially genetics) and how it will transform our definition of life. However, modern medicine also springs up delicate questions that are far from having an easy answer.

Some of the most controversial studies analyze embryonic stem cells. It is very likely that this research will provide the scientific answers to cloning, cancer treatment, and genetic medicine. In the current times, working with stem cells is difficult because of heavy regulations. Why? To *build* a stem cell, you need a live ovule and a spermatozoid. Once united, this cell will have the potential of transforming itself into any other cell in your body.

Some will argue that this is already the beginning of a child, and, therefore, we are committing a crime if we experiment with it. Some will say that it's just a cell that will start a mitosis[268] process and such a cell does not have feelings yet and, therefore, can be used for medical purposes. Others will argue that if stem cells have the potential to save millions of lives, then the sacrifice of a potential future kid is worth the boon we stand to gain. Without entering into the debate, it is clear that there is a controversy that hinders experimental testing. Where should the boundaries be? Is all the scientific improvement justified? Does the end justify the means?

This is just one example of a controversial topic that requires debate and legislation. There are many others in the medical world: abortion, child selection, GMOs in humans, the limits of disease detection, and the rights of surrogate mothers. It is clear that this industrial revolution will bring ethical trade-offs that we will have to address. There is even a term for these trade-offs: technoethics.

Technoethics is an interdisciplinary research area that draws on scientific theories and methods to provide insights about the ethical dimensions of technological systems and practices for advancing technological society.[269] [270] I couldn't insist more: there is a strong need for quick and effective legislation. Having fast, efficient, and global regulatory bodies for all technological challenges is mandatory. The risks of not having them could range from the outsourcing of medical research to countries with softer legislation to even human rights violations or unnecessary deaths. In other words, if technology is allowed to move too fast, it can become a severe problem for societies and, in the long term, a potential threat to human survival.

---

[268] Mitosis is a type of cell division that results in two daughter cells. Each one has the same number and kind of chromosomes as the parent nucleus, which is typical of healthy tissue growth.
[269] (Luppicini, 2010)
[270] (Jonas, 1979)

The trade-offs coming from freedom and privacy

Freedom is one of the most fundamental values in our democracies. However, as broad as the definition might sound, the use of technology poses specific ethical challenges to our freedom.

First, there is a clear trade-off between privacy and convenience. Let's illustrate with an example. Imagine that a new monitoring device appears on the market. You have to inject a small robot into your body, and it monitors some parameters such as heartbeat, glucose concentration, or blood pressure. This information will then be sent to the cloud, and you will be able to know how healthy you are and what to do to improve your health. Let's also assume that it is 100% safe and has zero secondary effects. Great product, right? Imagine now that you are an insurance company. Of course, you want to know how healthy your clients are in order to better evaluate their risks. With the information coming from this robot, you will be able to give higher prices to your unhealthy clients and vice versa. It might be a great promoter of a healthier lifestyle—and a great way to improve your profit margin. *Now* imagine that you're an employer. You'd like to pay less for your company's health insurance, so you decide to inject this robot in every employee to improve the overall health of your workforce. It might sound like a good idea, but the reality is that your employees might feel they are left with no choice. They will lose part of their freedom. Even if the employees stay healthier in the long run, they will be coerced into giving away part of their privacy and freedom to an external party. Guess what? Being unhealthy is also part of your freedom. Of course, this is just an example and change won't happen from one day to the next, but societies will have to decide what compromises concerning privacy they are willing to make and where they will draw a red line for these ethical issues.

Second, there is a clear trade-off between freedom and security. When Edward Snowden released the NSA's classified documents to the journalists of *The Guardian*, the world discovered that the NSA was directly spying on the whole internet, looking, supposedly, for terrorists hidden on the net. As you might expect, these spying actions were performed without the knowledge of civilians and through a secret executive order. After the publication of the NSA's files, the United States accused Snowden of being a national traitor, and the then-president of the United States, Barack Obama, opened the door to an ethical debate that should have been opened a long time ago. He clearly stated that to avoid a new 9/11, some sacrifices in terms of freedom had to be made. He was explaining that a public ethical debate had to take place to set the demarcation lines for NSA spies and surveillance in general. In other words, he stated that "you can't have 100% security and 100% privacy." This idea became known as the privacy trade-off. What is the price of freedom in the technological world? Is surveillance a must to maintain security? And can democracy survive without true privacy?[271] [272]

Environmental ethics

Last but not least, the environment will also play a role in the ethical challenges that come with technology. Technology has the potential to give you whatever you want here and now. It will change business models and enable companies to become global suppliers. However, this development has a cost, especially for the environment.

In the old times, trade was focused on the regional or national market. With the opening of global markets and the eruption of technologies that allow for supplying global demand, some ethical questions arise.

Let's look at the famous case of palm oil. Palm oil is a type of edible vegetable oil that comes from the fruit of the African oil palm tree. Oil palms are originally from Western Africa but can flourish wherever heat and rainfall are abundant. Today, palm oil is grown throughout Africa, Asia, North America, and South America, with 85% of all palm oil globally produced and

[271] (Poitras, 2014)
[272] (Obama, 2013)

exported from Indonesia and Malaysia. Palm oil is everywhere. Palm oil and palm-kernel-oil-based ingredients are found in almost 50% of products on supermarket shelves, including food and non-food items.[273] It's in everything, from your Nutella jar to your favorite shampoo. Palm oil has a high resistance to oxidation and, therefore, has a long shelf life. Because we can technologically afford to reach the globe, the production of this oil is increasing dramatically and having a massive impact on the environment. Its expansion occurs at the expense of biodiversity and native ecosystems in the countries where it is produced. Just as an example, one-third of all mammal species in Indonesia are at critical risk of extinction as a consequence of unsustainable development that is rapidly encroaching on their habitat. Deforestation for palm oil production also contributes significantly to climate change. The removal of native forests often involves the burning of invaluable timber and forest undergrowth, emitting immense quantities of smoke and greenhouse gases into the atmosphere.[274] [275]

Another issue linked to technology is the concept of "planned obsolescence." Because society is now capable of producing more and more efficiently, a new need appears: products have to have shorter lifespans to maximize profitability. In a nutshell, companies purposefully make products last for less time so they can make more money by selling more products. This behavior is not new (see the Phoebus Cartel in Box 2), but the global production systems derived from technology enhance it. More and more companies are becoming Fast Mover Consumer Goods corporations (also called FMCGs). Even Apple admitted that older iPhone models were deliberately slowed down through software updates. Rotation of products is financially vital even if it implies environmental damage. The problem with technology is that it has given us the tools to make this FMCG mindset work with all the goods in the market. These behaviors are unsustainable for the environment (higher need for raw material,

---

[273] (Green Palm Sustainability, 2017)
[274] (Say no to palm oil, 2017)
[275] (Greenpeace, 2016)

more emissions, more waste, among others) and are based on questionable ethics.[276] [277]

Final thoughts

In summary, the ethical frameworks that societies have will need to be adaptive enough to accommodate the fast challenges of technology. Those complex ethical frameworks will require quick legislation and a rapid understanding of the risks involved. If technology can move faster than our knowledge of it, we risk losing control of it. When we want to regain control, it will be too late. We have to be able to leave space for innovation while maintaining the correct global security for humanity. Not an easy task, especially with the challenges that come from artificial intelligence or genetic engineering. Ethics will have to define the relationship that we want to have with technology and the level of development that we want to reach. We are not machines. We are humans.

Box 1: The Deepwater Horizon oil spill

The Deepwater Horizon was a 10-year-old floating drilling rig that could operate in waters up to 10,000 feet (3,000 meters) deep. The well was located in the Macondo Prospect in the Gulf of Mexico. At approximately 9:45 p.m. on 20 April, 2010, the high-pressure methane gas from the well expanded and exploded, causing the largest marine oil spill in the history of the petroleum industry. The US government estimated the total discharge at 4.9 million barrels (780,000 m^3) of oil. Eleven people died and 19 were injured. As of February 2013, criminal and civil settlements had cost the company $42.2 billion. In September 2014, a US District Court judge ruled that BP was primarily responsible for the oil spill. The company agreed to pay $18.7 billion in fines, the largest corporate settlement in US history.[278] [279]

---

[276] (BBC, 2018)
[277] (BBC, 2016)
[278] (Deepwater, 2010)
[279] (Wikipedia, 2018)

This oil spill is just another example of the environmental costs of worldwide development. Zero risk doesn't exist and never will. The impact that our industries are having on the planet should make us think about the risks we are comfortable with and which are justified to fuel our progress. Clearly, if we continue to drill huge quantities of petroleum in the sea, we will have another oil spill in the upcoming years. It is clear too that without oil, we will not be able to maintain our rhythm of development. Agreeing upon the ethical limits to allow such operations is essential to progress.

Box 2: The Phoebus Cartel

Light bulbs are probably among the most emblematic case studies of planned obsolescence. Thomas Edison invented a commercially viable light bulb in 1880. Early adopter consumers were not keen on paying for replacement units of this new technology, so the companies produced bulbs to last. However, this reality changed when light bulbs reached the mass market around 1920. Producers realized that more business could be done if the light bulbs were disposable. To achieve this goal, the leading manufacturers of the moment (Osram, Associated Electrical Industries, and General Electric) agreed to artificially reduce the bulbs' lifetime to 1,000 hours. This secret plan would later be discovered and named "the Phoebus Cartel."[280]

Programmed obsolescence is present in our current times. Almost all consumer goods have a foreseen lifespan that has been defined artificially by the manufacturers themselves. However, it's overly simplistic to condemn the practice as entirely wrong. Frequently, planned obsolescence benefits both the consumer and the manufacturer. Let's take the classic example of smartphones. Many owners might appreciate paying less for a phone that has a planned lifespan of three years. In three years, the technology will have evolved so much that the phone that you once bought is no longer useful. Technology is rendering itself

---

[280] A cartel is an association of manufacturers or suppliers that is established with the purpose of maintaining prices at a high level and/or restricting competition.

obsolete every year with the innovations that reach the market. The other side of the coin is that this programmed obsolescence has a cost to the environment that is almost never included in the equation. Agreeing on the correct balance will be delicate.[281]

Box 3: Not enough languages in the world?

Facebook abandoned an experiment with two artificially intelligent programs that started to talk to each other. The reality is that the machines were so successful that they created a language from scratch by themselves that wasn't understandable to the programmers. At the end of the experiment, the scientists decided to power off the machines. We are currently at an early stage in machine learning technology. If we are now facing problems like indecipherable robot languages, imagine what machine-learning robots will be able to do in the future if we are not careful with them or if they fall into the wrong hands. Are we setting the correct barriers?[282]

---

[281] (BBC, 2016)
[282] (Independent, 2017)

In the future, society may become a scary and unfamiliar place to live. Indeed, we are facing changing times, and we have our feet in quicksand. In a fully interconnected world where privacy is limited, we will fail to fully develop our personality and identity without the fear of being observed (and judged). At the same time, technology will begin to solve problems that are not human enough to be understood. Part of society might be left behind and, who knows, maybe we will be part of those left behind one day. Is your digital footprint safe? Are you comfortable with the current speed of technology? Is it so fast that you see people being left behind? Is it so slow that you don't have the framework to accommodate this new reality? Are you part of the problem or the solution?

C. How will **you** be afraid of yourself?

Technology will make the world a smaller place. We will be able to connect and to instantly share who we are with people on the opposite side of the world. We will be able to cheaply trade with a country that has low wage labor, low taxes, or both, no matter where it is geographically located. Like it or not, technology will directly change who we are. You will be able to be who you want to be (and not who you are). However, we are social animals that can have a maximum of 150 stable relationships.[283] Let's be clear: even if technology is changing you, it will not change our species. The age of massive IT connections will dilute our human empathy. We will lose our connection with our peers and with the motherland that gave birth to us. We run the risk of realizing that we don't know who we are anymore. We will recognize that too much technology can reduce us to the zeros and ones that we once trusted.

i. Failure to accept human imperfection
To begin with

Technology has shaped generations of human development since the beginning of humanity. The discovery of fire and the ability to control it utterly changed the organization of societies and the definition of power, control, and purpose. We changed our way of seeing the world. Aren't we in a similar new paradigm? What's different this time? Aren't we just communicating, sharing, and behaving as before but in a digital world?

Human communication

First of all, we are shifting the attention we once gave to people into those little devices that we carry 24/7 in our pockets. Parents don't take care of their kids anymore during breakfast; they are too busy replying to their emails and WhatsApp messages. What's worse, kids don't pay attention to their peers during breaks; they are too busy retouching their latest Instagram post

---

[283] (Purves, 2007)

and checking how many likes they have gathered. The statistics show that the average American older than 18 years of age spends over four hours on their smartphone and/or tablet every day.[284] And this time is taken out from one of our core activities: human interaction. It is a paradox. We are entering a world where we are always connected with everyone but never fully bonded with anyone. If we think about it, it is no surprise that this phenomenon occurs. Face-to-face communication has lots of *problems*; it happens in real time, you can't control what you are going to say, and things can quickly go wrong. However, the worst part is that you have to project the real, true image of yourself, with all your imperfections that make you who you really are.

Technology, on the other hand, lets us present ourselves as we want to be seen. We are able to edit, retouch, and even delete parts of ourselves that we don't like. We control our virtual lives, but we can't always control our physical world. Guess what? Human relations and real human trust require those imperfections and vulnerabilities to exist. We sacrifice conversations for mere connections that will never truly fulfill us.

To make matters worse, this virus called technology is infecting younger and younger generations. One day, they will realize that they don't know how to have a genuine and meaningful conversation with the people they call friends. It seems that we are forgetting that our happiness and fulfillment are directly linked to the people who surround us and interact with us. And believe it or not, Siri or Alexa will never replace human beings, even if, unlike your friends, they are always there to listen and reply to you. But those robots programmed with artificial intelligence are not listening; they are just recording your voice. They don't care for you; they are just following an algorithm. They don't feel. They are just telling you what another human told them to. Machines will never have the most important human attribute of all: empathy. They are just faking it, and we are aware

---

[284] (Hackernoon, 2017)

of that. Human connection in real life is what makes us human. We are and will always be social animals.[285]

A sense of imperfection and the wabi-sabi concept

We always worship the best among us in every aspect of our lives: best athletes, best managers, best employees, best actors, etc. We never idolize the person who ended fifth in a race. Moreover, we need public recognition of our winners. That's why we have prizes, gold medals, red carpets, Fortune lists, etc. We have an innate collective appetite for awesomeness. Of course, it makes sense. We, humans, want to thrive in life. To do so, we want to develop ourselves by emulating our heroes. And technology allows us to do that.

Technology is always young, beautiful, and cool. Ever perfect, still the best. It allows us to choose the image of ourselves that we want to project to the whole world. This editing power brings a cost: the virtual profile we create of ourselves rarely represents our reality. It is unlikely that anyone would post on Facebook how bad their holidays were or how they failed in a relationship or that one picture taken at 7 a.m. just after waking up. We all know that in life there are good and bad moments. However, on your Instagram profile, you will only share what shines in your life (or you will fake it). This faking of experiences brings us to a point where we all know that everything we see on social media has at least been modified and is far from reality. The famous "I think, therefore I am" by Rene Descartes is replaced by an unfulfilling "I share, therefore I am." Never forget that what makes us human is that we are far from being perfect. That's what trust is built upon.[286]

There is a very enlightening concept in the Japanese culture called wabi-sabi. Inspired by Buddhist teachings, it recognizes asymmetry, irregularity, and modesty as attributes of beauty. Wabi-sabi represents the acceptance of imperfection. In other

---

[285] (TED, 2012)
[286] (TED, 2012)

words, wabi-sabi is the art of accepting that imperfection makes us great. It embraces all that is authentic by acknowledging three simple realities: nothing lasts, nothing is finished, and nothing is perfect. For example, the wabi-sabi sense of beauty is based on the appreciation of the process of aging. It wouldn't make sense to try to stop our natural progression through life. All our heroes are also imperfect, even if they apply a filter to their Instagram posts. If we want to strive for the best, we will also need to accept our imperfections and share them. It is only then that we will become real and inspiring human heroes. [287] [288]

Cyberbullying's impact on young people's mental health

Social media is a common place for most young people to share their opinions, interact with each other, and access endless information. However, it also presents risks that are sometimes forgotten or hidden behind the shine of the platform. One risk is online harassment (also called cyberbullying). In the old times, school grounds were the field for bullying. However, with the 24/7 access that teenagers have to the digital world, harassment and abuse have entered young people's homes. Bullying can happen at any time, which creates constant stress and anxiety for the victims. Younger generations are the most vulnerable to cyberbullying, and the impact on their mental health is already a reality.

Let's take a look at the statistics for young people (aged 11 to 25):

- 44% spend more than three hours a day on social media.
- 38% reported that social media has a negative impact on how they feel about themselves, compared to 23% who said that it has a positive impact. This percentage is more critical for girls, 46% of whom stated that social media had an adverse effect on their self-esteem.
- 83% are calling for social media companies to do more to tackle cyberbullying on their platforms.

---

[287] (Fast Company, 2013)
[288] (Powell, 2004)

- 46% wouldn't tell their parents if they had a bad experience on social media.
- 47% have received intimidating, threatening, or nasty messages online.

There is also scientific evidence that children who were victims of bullying are much more likely to have low subjective well-being than other children. Quantity also matters. Frequent bullying increases the likelihood of victims having low well-being. Being bullied has also been associated with symptoms of mental illnesses like depression, and the scars of that trauma can last into adulthood. Furthermore, children and young people who become victims of cyberbullying are more than twice as likely to indulge in self-harm and attempt suicide than are non-victims. Research doesn't sound too optimistic and shows that the problem is widespread. Of course, bullying has always existed (even before the internet), but the fact that people are now just a click away from being "connected" 100% of the time gives a new dimension to the problem, which makes the potential solution even more complicated.

It is our collective responsibility as a society to fight against online harassment. First, social media companies must be age appropriate, allow parental control, and require explicit parental consent for younger users. In addition, they must provide timely, effective, and consistent responses to online bullying and be held accountable legally if they don't. Social media companies should be required to publish data about their reaction to reports of online bullying. Schools should teach children and young people how to be safe and responsible online and ensure they know how to respond positively to online attacks such as cyberbullying. It is also essential for the teachers themselves to be trained in new technologies, even if they belong to other generations. They must adapt to the new reality. Last but not least, parents play a crucial role to solving this problem. They have to understand the risks of social media and have an open debate with their kids about it. Moreover, in case of a problem, they must know how to react. Cyberbullying is a national health hazard that needs to be tackled, and it is our societal duty to do so.[289]

Final thoughts

Social media, and technology in general, is probably the best tool that humanity has created to connect with each other and make the world a smaller place. However, it has also made our lives fake and empty by hiding our true, imperfect nature behind its facade. We are social animals that need true and imperfect physical connection. That's how trust emerges. That's where our real potential as human beings appears. We will need to change our relationship with technology if we want to live in a better place than before.

Box 1: Loneliness increases the risk of premature death

The percentage of Americans who report chronic feelings of loneliness has risen over the past few decades. The percentage of Americans who admitted that they often felt lonely was between 11% and 20% in the 1970s and 1980s (depending on which study you take into consideration). However, in 2010, the American Association of Retired Persons (AARP) did a nationally representative survey and found that the percentage of people with chronic loneliness was closer to 40 to 45%.[290] And loneliness is not only found in older people. According to a study from the General Social Survey (GSS), the number of American citizens who say they have no close friends has roughly tripled in recent decades. When people are asked how many confidants they have, the most common response is zero. Loneliness particularly affects adult men, who seem to be especially underperforming at maintaining and cultivating friendships.

All this may seem strange in the era of Facebook, Twitter, and great digital connectivity. But the "friends" orbiting us in our digital galaxy aren't the ones that matter when it comes to our health and happiness.[291] An NHS study analyzed 70 independent

---

[289] (The Children Society, 2018)
[290] (Fortune, 2016)
[291] (Time, 2015)

prospective studies involving more than 3.4 million participants who were surveyed for an average of seven years. Overall, the researchers found social isolation resulted in a higher likelihood of death, whether measured objectively or subjectively. NHS data showed that the increased risk of death was 26% for reported loneliness, 29% for social isolation, and 32% for those living alone.[292] We will need to remember these risks if we want to maintain a healthy relationship with technology.

Box 2: Coping with being alone

With technology, people are always connected. It has permanently opened a window to share everything we want. We are never entirely alone. When we feel lonely, we send a message to our contacts to receive their replies and feel loved again. We are starting to lose our ability to be alone.

A teenager who has always had technology at hand has never been fully alone. The problem is that being alone shouldn't be a problem in and of itself. Being alone is a skill to learn, especially in early ages, but new generations are growing up without knowing how to comfort and entertain themselves when they're alone. Being alone is not something bad that has to be eliminated from our lives! It doesn't have to be boring. The privacy that being by yourself brings, with no distractions, gives you the chance to clear your mind, focus, and think more lucidly. It's an opportunity to revitalize your mind and body at the same time. Constant motion prevents you from engaging in deep thought, which inhibits creativity and lessens productivity. Being alone from time to time is cool and healthy.

Box 3: Banning social media at home?

Kowhai Intermediate School in Auckland (Australia) has appealed to parents to ban social media for the entire two years that their kids are enrolled. The objective is clear: the school wants to fight inappropriate digital behavior following an explosion of social

---

[292] (National Health Service UK, 2015)

media and online bullying. Access to social media is already banned during school hours and on school grounds, but this is the first time parents have been asked to prevent access at home too. This measure has brought some controversy with different opinions on the table. On one side, those who support the idea explain that schools are not strong enough to protect their students from cyberbullying and that social media platforms have too much influence. On the other side, the detractors believe that banning social media is too drastic and that kids are losing the opportunities that those platforms bring (from a sense of belonging to a new type of socialization). The debate doesn't seem to be wrapping up anytime soon.[293]

---

[293] (New Zeland Herald, 2018)

ii. Lost connection with Mother Nature
To begin with

The environment has been long forgotten by governments and citizens. All industrial revolutions have affected our planet's equilibrium. However, this last one will be far more aggressive than ever before. Humankind now has access to advanced machinery that can make a mountain disappear in a few months or change the fauna in a region within weeks.[294] Our planet is not ready for all these quick changes. Technology has made us lose our connection with the Mother Earth that gives us everything we have. Globalization and its needs will completely change the delicate ecosystem that we live in. Some countries are making significant efforts toward cleaner and more sustainable development. However, the developing countries that are joining the league of developed nations have a strong argument: "it is now our turn to develop ourselves, even if we contaminate the earth, just as you did in the last industrial revolutions." Living in this new world with so many players wanting to preserve their growth at any cost will be complex.

Cities will be the new normal

If you are looking for opportunities, you must move to a city. Cities and mega-cities are the backbones of this new industrial revolution. Cities do and will attract talent from all over the world. The majority of innovation will happen in cities. Indeed, if you are looking to boost your professional career, a move to a big city will definitely increase your chances of getting a higher salary and increasing your purchasing power.[295] Needless to say, entrepreneurship levels are higher in cities just because of the size (more people means more stuff happening). Ideas come from sharing different points of view. However, this move might also have a setback. Living in a city might make you forget the roots of your origin. As a reminder, we all come from Mother Nature. Kids in large cities are growing up without ever having

---

[294] (Climate Central, 2013)
[295] (Canadian Government, 2010)

seen the stars. We build bigger cities, but we're not aware how much and how fast we're undermining our connection to nature. This can directly affect our health. Numerous studies have suggested that urban living conditions undermine mental health, whereas conditions in rural areas support it. There are numerous environmental factors that can increase stress in urban settings and, thus, the prevalence of disorders such as anxiety and depression is on the rise. If cities don't add natural features to their layout, they can simply become a risk to our health and psychological stability.[296] As extreme as it may sound, most of the most common causes of death are directly or indirectly influenced by faulty urban design and planning policies. For example, heart attack (#1 cause of death), stroke (#2), chronic respiratory disease (#4), and lung cancers (#5) are between 25 and 33% caused by air pollution, which comes mainly from urban traffic, waste, industry, cooking, heating, and power production. Pneumonia (#3) is also caused by air pollution more than 50% of the time. And diabetes (#6) is linked to obesity and physical inactivity, which is common in car-dependent cities. Cities produce 75% of carbon emissions, while urban populations are among the most vulnerable to climate change. Moreover, direct climate change costs to health are expected to reach $ 2-4 billion annually by 2030. Cities are costing us money yet eventually killing us. A health-focused urban design will be essential to create a connection between cities and humans again and to remind us that we are connected to our world.[297]

Mass production emissions

Let's be clear. Mass production will produce mass emissions and mass waste. It is true that technology will help us to monitor and reduce these emissions. But size matters. Over-consumption of material *things* that we don't need is affecting our planet. Even if we have short-term gains from mass production, there is an ecological bill that is not taken into account. The convenience of this revolution is making us forget what the true cost of fashion,

---

[296] (Terry Hartig, 2016)
[297] (World Health Organization, 2018)

food, cars, and houses is regarding emissions, lake contamination, water consumption, communities' destruction, or minerals. Just to give an example, the production of a typical two-gram computer chip takes 1.6 kilograms of fossil fuel, 72 grams of chemicals, and 32 liters of water. Just for two grams![298] Those chips are in all the electronic devices in the world. Another example: producing one kilogram of beef requires 15,000 liters of water. Multiply that by all the beef that you eat yearly and you can have an estimation of how much *virtual* water you consume per year only by eating this kind of meat. An average meat consumer in the United States with a yearly income of $40,000 will consume in average 2,600 m^3 of water per year. In other words: a 14-cubic-meter swimming pool.[299]

Hopefully, technology and an open internet will help us monitor our suppliers so they will at least be accountable for their actions. However, the fact is that even if efficiency increases, our current rate of production in absolute terms is too high and therefore unsustainable. Even more, the world's population is expected to rise to almost 10 billion, with an increasing middle class that will be able to afford more stuff.[300] Capitalism teaches us that growth is what brings development, but we have to be aware of the costs that development brings to our planet. If we continue down the same path, we will face dramatic consequences regarding greenhouse emissions, climate change, resource consumption, climate refugees, and climate disasters.

Where does anything come from?

In the old times, food was sold in markets and was harvested from the farms next to your city, and you could only find seasonal products that were cultivated by the person trying to sell them. Globalization and the emergence of cheap transportation and technology triggered a shift. We are now able to buy avocados from Peru, bananas from the Caribbean, and grapefruits from

---

[298] (Eric D. Williams, 2002)
[299] (Water footprint, 2005)
[300] (The Guardian, 2015)

China in our supermarkets. Those exotic places have brought variety to our tables, but they have also made us less aware of where our food comes from, not to mention the environmental impact of such global supply. It is challenging to track down the origins of any item we purchase in any store anymore. Therefore, it is also difficult to keep the companies that are, for example, using chemical products that we may not want in our dishes, accountable. The health effects of fertilizers and pesticides are still being researched, and we will need more time to identify the impact that they have on our health. However, the reality is that technology makes production processes invisible to the end consumer. Of course, companies take advantage of this invisibility. Hidden supply chains are a reality for food—and for all other goods. Unfortunately, there is a long list of examples. Nike has children as young as 10 making shoes, clothing, and footballs in Pakistan and Cambodia.[301] Apple's assembly-line workers began committing suicide in Longhua (China) due to high stress and long hours.[302] PepsiCo, Unilever, and Nestlé were accused of complicity in illegal rainforest destruction in Sumatran elephants' habitat in Indonesia.[303] It is our lack of information, our classic human *laissez-faire* attitude, and the opacity of the system itself that permits all these hidden costs. We are losing the connection with Mother Earth and with our values. I am optimistic that technology might also be the solution. However, we are far from there, and it is the responsibility of governments and consumers to fight for a more transparent and environmentally friendly management of supply chains.

Final thoughts

Clearly, this industrial revolution has made us lose our connection with the planet that sustains our life. The shiny brightness of technology has erased this link. Cities are the keystones for human progress in a new capitalistic world, even if they make us forget the smell of fresh grass. Our overconsumption fulfills all

---

[301] (The Guardian, 2006)
[302] (The Guardian, 2017)
[303] (The Guardian, 2017)

our needs—as long as we forget their real costs. Last but not least, our need for variety has brought us worldwide supply without taking care of the ecological footprint that it conveys. The future is uncertain, but we need to make an effort to move toward a more efficient and transparent way of doing business. Without a change in mindset, it might soon be too late.

Box 1: Lagos as an example of a future megacity

Africa's population is expected to double by 2050, reaching 2.5 billion people. One example of this population explosion is Nigeria's largest city: Lagos. The UN estimated the population at 12.6 million in 2014 (higher than, for example, the population of Belgium, which is around 11 million). The city is growing without breath; more than 500,000 people move to the city every year. As you might expect, some challenges come with this immigration. One of the most critical challenges is access to water. Estimates suggest that only one in 10 people have access to water supplied by the state utility provider. Despite being Africa's largest economy, Nigeria has one of the highest child mortality rates from water-borne diseases in Africa. The new megacities that are planning to become the megalopolises of the future will need to be able to adapt fast and invest efficiently if they want to sustain themselves without being drowned by their success.[304]

Box 2: Paris Agreements?

The Paris Agreement is an agreement within the United Nations Framework Convention on Climate Change dealing with the mitigation of greenhouse gas emissions. The agreement was negotiated by representatives of 196 parties and adopted by consensus on December 12, 2015. The agreement aims to repond to the global climate change threat by limiting the global temperature rise this century to well below two degrees Celsius above pre-industrial levels. If successful, this would bring massive relief. As I am writing (January 2018), 174 states and the EU, representing around 87% of global greenhouse gas emissions,

---

[304] (Reuters, 2016)

have ratified or will soon ratify the agreement. The list includes China, the United States, and India, three of the four countries with the most significant greenhouse gas emissions. However, on June 1, 2017, US President Donald Trump announced that the United States would withdraw from the agreement shortly. Again, we have here an example that shows that we still tend to give a higher priority to our national economic agenda than to global problems such as climate change. Without a global engagement, we will not be able to stop the destruction of our planet.[305]

Box 3: The nature conservancy

Two out of every three people will call cities home by the middle of the twenty-first century. This historic urban growth, coupled with a changing climate, invites organizations to team up with communities to ensure a future where nature and people thrive in cities. One example is the NGO, the Nature Conservancy. Their mission is to conserve the lands and waters on which all life depends. They are developing strategies to ensure a healthier quality of life for people in cities by making them more natural. They build initiatives with local NGOs and municipalities in each town of the United States. For example, they develop gardens and local farms in the schools of the city to make kids (and parents) more aware of the food production chain.[306] These initiatives are the ones that bring hope to humanity by accepting that development has to include nature preservation in its equation.

---

[305] (United Nations, 2018)
[306] (The Nature Conservancy, 2018)

iii. Lost pride in being a human being
To begin with

Abraham Maslow, one the most prominent psychologists of the twentieth century, stated that humans are motivated to fulfill certain needs and that some of them take precedence over others. Our most basic need, logically, is physical survival, and this is the first thing that motivates our behavior. Once that need is fulfilled, the next level up is what motivates us, and so on. Maslow described five steps in his pyramid to describe the pattern that human motivations generally move through. They are (in order): physiological needs, safety, belonging and love, esteem, and self-actualization. However, it seems that the shine from technology is darkening our core definition of what a human being is. If machines are going to master the world, what is the meaning of life? What is the value of a human being? Have we just built our tomb?

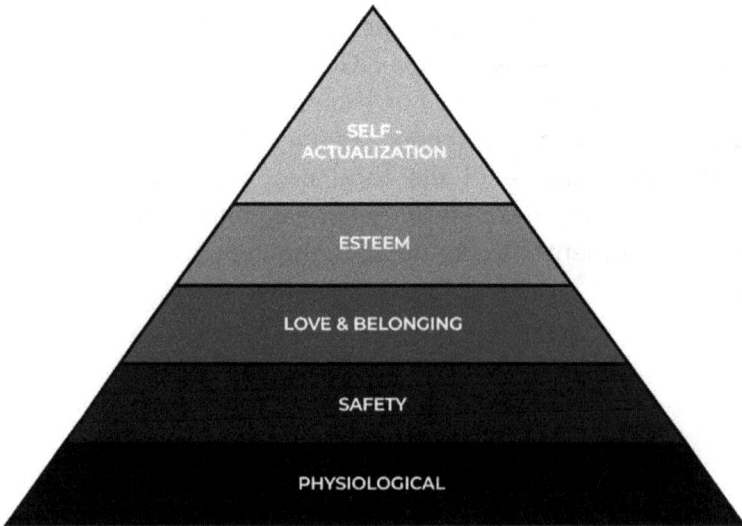

FIGURE 1: PYRAMID SHOWING MASLOW'S HIERARCHY OF NEEDS[307]

---

[307] (FireflySixtySeven, 2014)

The useless class and the quest for purpose

Work is an important part of our lives and our pride. If your average working week accounts for 40 hours, you will spend around one-fourth of your life working. That's a lot of time. Work brings purpose to our day-to-day existence. As technology renders jobs obsolete, what will keep us busy? That is one of the most common fears that we all have about technology. The majority of jobs that exist today may disappear within decades because of automation and artificial intelligence. Even if new jobs are created, they will not be enough to cover the loss. Not only will we have fewer possible jobs, but there will be a new class emerging: the useless class.[308] It might sound a bit nasty to call them that, but the reality behind this term is much more profound that you may think. First of all, the word "useless" doesn't mean "meaningless" in this context. Those individuals will have a huge sense of meaning, but they will be outside of the workforce. A person would be part of the "useless class" if she or he was unemployed and unemployable. If work is not a part of our lives, a new sense of purpose will need to be built if we want to avoid mental depression and social unrest among other problems. There are a lot of different ideas that have come into place to solve this crisis, ranging from a universal basic income to a complete change in the education system. However, it seems that the only real solution will be a mix of all of them.

On the other side, technology will probably also bring solutions in the quest for purpose, especially when artificial intelligence is at a higher level. Augmented and virtual reality powered by artificial intelligence will provide the people who have lost their sense of purpose with a reason to live happy and meaningful lives. The historian Yuval Noah Harari describes AI as the religion of the future. Those virtual realities are likely to become the key to providing meaning to the useless class of the post-work world. We still don't know how the process will look. It may involve 3D glasses, sophisticated smartphones, new religions, or new gurus

---

[308] This concept of "useless class" has been created and studied by the historian Yuval Noah Harari.

with new ideologies. The fact is that this reality is already on the streets; for some people, virtual life is already more important than real life. When the new fiction becomes the new reality, I hope you will be able to remember the smell and feeling of human interaction.[309] What is clear and factual is that technology will affect our sense of purpose and have a profound impact on our lives. Users' happiness will be an intrinsic characteristic of the future technological products that will appear on the market. However, the question remains: is technology the answer to human purpose? The answer is a resounding "yes, it is." Time will tell to what extent.[310] [311]

The MeMeMe Generation

Let's be clear. The new industrial revolution has brought with it a new generation called "millennials" (born between 1980 and 2000). They are the children of the baby boomers and have grown up with technology surrounding them. Their parents call them lazy, entitled, selfish, and shallow (just like any generation). However, this time, data supports those statements.

Millennials are true believers that they are unique and amazing. The incidence of narcissistic personality disorder is nearly three times as high for people in their 20s as for the generation that's now 65. Forty percent believe they should be promoted every two years, regardless of their performance. Narcissism has been labeled a "modern epidemic," pointing to the rapid change in society that occurred in industrial and post-industrial times. The past few decades reflect a societal shift from a commitment to the collective to a focus on the individual or the self. Educators and parents told their children how special and unique they were to make them feel more confident. However, technology gave those children the tools to express their exceptionality to the world—a world that was always listening or, at least, a world that always

---

[309] Even if it might sound a bit controversial, it is important to note that this section is not intended to judge whether having a digital life is a good or bad thing to happen. I am not advocating for either side.
[310] (The Guardian, 2017)
[311] (Harari, 2015)

seemed to be listening. Parents tried to confer self-esteem upon their children, rather than letting them achieve it through hard work. As the old social fabric was falling apart, it became much harder to find one of our most basic needs: meaningful connection. The modernization of society prized fame, wealth, and celebrity above everything else.

It is even more interesting to realize that because of globalization, social media, the influence of Western culture, and the speed of change, millennials worldwide are more similar to one another than they are to older generations within their own nations. This trend is not only affecting the wealthy families. It affects everyone. As surprising as it may seem, poor millennials have even higher rates of narcissism, materialism, and technology addiction. The problem is global and will only be worsened by the current state of globalization. If human connections, community values, and humbleness are not promoted, we risk losing the vision of who we really are, of what our true self-worth is.[312] [313]

There is only one global culture

In countries outside the United States, the influence that American culture and business has on a country is termed "Americanization." Their media, cuisine, business practices, popular culture, technology, or political techniques can all become "Americanized." Americanization has become more prevalent since the collapse of the Soviet Union in 1989-91 and especially since the internet reached critical mass in the early 2000s.

The globalized world has become a centralized place in the cultural aspect. Companies have not only launched new products but also promoted a particular lifestyle. There is no doubt that one culture has outperformed them all: the American culture. Brands such as Starbucks, Coca-Cola, Pepsi, McDonald's, Burger King, Pizza Hut, Kentucky Fried Chicken, Domino's Pizza,

---

[312] (Time, 2013)
[313] (Independent, 2016)

Microsoft, Apple, Intel, HP, Dell, IBM, Google, Facebook, Twitter, Uber, Disney, Pixar, DreamWorks, Illumination, HBO, Netflix, Universal, Century Fox, etc., can be found in any major city in the world. No other country can pride itself on having such a worldwide economic and cultural influence.

The problem is that the international rise of the American culture is superseding the local cultures that countries had, even to the point of erasing some of them. As you might expect, there have been movements against this economic and cultural expansion. Far-right parties outside of the United States have advocated for closing their borders and global trade, and other countries have created federal incentives that focus on keeping the local culture and companies alive.

However the future may look, the reality is that this industrial revolution is moving societies to a more unified and global culture, which might be good and bad at the same time. Currently, a kid in Japan has lots of things in common with a kid in Spain. They will probably have a smartphone, frequently update their Facebook profile, drink Coca-Cola from time to time, and know who Justin Bieber is. This reality might be a sword of Damocles. On one side, the world will become a smaller place, and interaction and innovation will rise from it. On the other side, it will be more and more difficult to belong to a *unique* culture as a community that shares the same group of values and beliefs. The risk of diluting this group might make us realize that we are neither unique nor special anymore.[314]

Final thoughts

Philosophy and psychology books need to be updated every year. New realities require a whole new thinking paradigm that wasn't foreseen before. A whole new range of human behaviors have to be analyzed to be understood. Maslow described a world where a five-layer pyramid of needs drove human motivation. However, it seems that definitions are changing. The lack of

---

[314] (The Guardian, 2014)

purpose due to a new post-work world, the discovery of new types of goals thanks to technology and artificial intelligence, a society that is globally connected but that has somehow lost the true meaning of brotherhood, and the globalization of culture will change the rules of the game and bring us deep into unknown waters. Nobody knows where we will finally land in the next decades to come. However, the question remains. Will we be proud of and comfortable with our new purpose? Will technology be the problem, the solution, or both? Will we pay for human convenience by selling human pride?

Box 1: Lack of things to fight for

Heroes and visions bring hope to societies. However, in the developed world, it seems that the industrial revolution we are building is erasing some of the grand dreams that used to drive our parents to get up every morning and go to work. The new generations are exchanging the trust in human connection for the excitement of social media and chatting platforms. They are exchanging work for automation. The countries that we once fought for are failing at finding their place in the world and at making people dream about them. Even love seems to be a *swipe right* far from us.

The things we once fought for (connection, work, nations, love) are being diluted by technology. And it may one day be too late to change it.[315]

Box 2: The EU and the film industry

The strong presence of Hollywood productions characterizes the EU film landscape. In 2013, Hollywood held a share of nearly 70% of the EU market, while European productions represented only 26%. What makes the major US companies so powerful, in the words of the European Parliament, "is the fact that they are vertically integrated, with activities spanning production and distribution, allowing them to spread risks over several films, and

---

[315] (Inés Figueiredo, 2017)

reinvest profits in new projects." To balance the financing challenges facing EU film companies, different types of film-support schemes have been set up, accounting in 2009 for around €2.1 billion (excluding tax incentives and interventions by publicly funded banks and credit institutions).[316] Therefore, like it or not, the EU countries have gathered together to fight against the US influence by directly funding EU film projects. It might sound like a clear example of disloyal competition. The debate is served.

Box 3: Millennials and the establishment

The relationship between the establishment and the millennials is a love/hate story. The fun part is that the social revolution is not taking place because millennials are trying to take over the establishment but because they're growing up without one. They claim that they don't need one. The industrial revolution has made individuals far more powerful now than they have ever been in history. People can move freely and cheaply, start a business, access global knowledge, make international transfers, and form organizations easily. The information revolution has further empowered individuals by handing them the technology to compete against huge organizations: hackers vs. corporations, bloggers vs. newspapers, terrorists vs. nation-states, YouTube directors vs. studios, app-makers vs. entire industries. Millennials don't need the establishment. That's why the establishment is fiercely scared of them.[317]

---

[316] (European Parliament, 2014)
[317] (Time, 2013)

We haven't transformed our species, but we have changed the world we are living in. Somehow, we might be afraid of the transformation that we have enjoyed/suffered/built during such a short period. Our pursuit of perfection has erased the old values and attitudes that made us great in the first place. We have lost part of the empathy that made us human and a substantial portion of this loss comes from technology. Human beings are, by definition, always afraid of change, and these times are no exception. Have we already drowned in the zeros and ones that we once trusted? Have we already gone too far? Will we be able to live in this new reality?

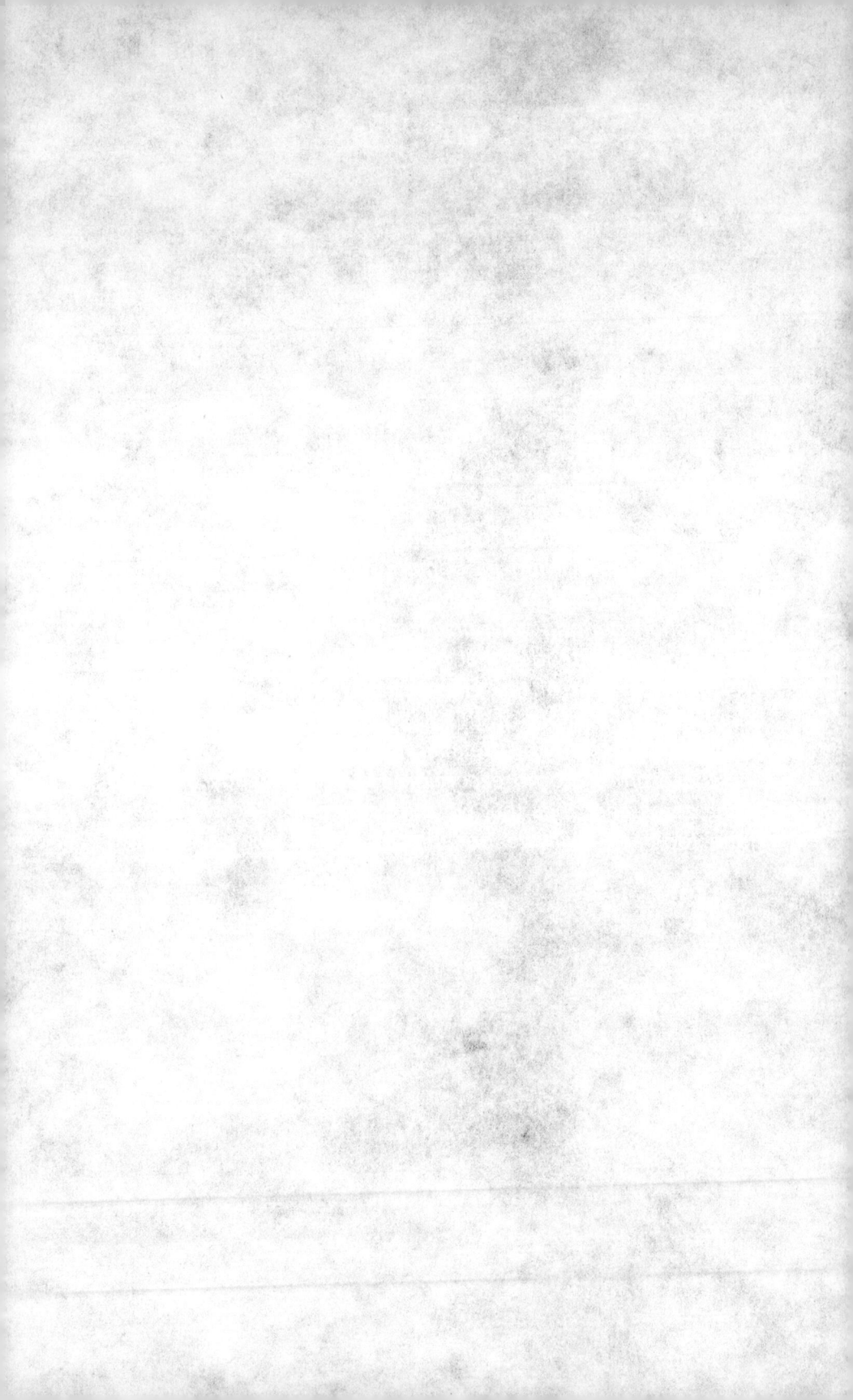

As a summary, we have seen that this new industrial revolution has many setbacks that are sometimes forgotten (perhaps intentionally, perhaps not) while we marvel at its shine. A lot of technological advancements are already on the market, and the required regulation has, as you expect, not yet arrived. Furthermore, our technological mindset is neither ready nor educated to face the challenges that come from our innovations. The impressive part is that, somehow, the companies surfing on the wave of this industrial revolution have such a good reputation that they are sometimes considered as untouchable as Gods by their consumers. We will one day have to honestly face the downsides of technology and make the hard decisions. We will have to choose between privacy and convenience, Planet Earth and mass production, and human interaction and Instagram likes, among many others.

Scary, huh?

The scary part is not that trade-offs have to be made when innovations enter the market. What is terrifying is that the public is still ignorant about the hidden costs of technology that may have a massive impact on our future lives. On the positive side, if you've read everything we discussed so far, you are now already aware of the main setbacks that we are facing. And well, the first step to solve a problem is to be aware of it, right?

We've seen that innovation is speeding up. The trade-offs from it are the new black, and we will have to deal with them.

You want to have it all? Get ready to be adaptive and fast. **Period.**

You want to become Superman? Get ready to assume the costs of technology. **Period.**

You want to feel unique? Get ready to shake off your ethics. **Period.**

"And they all woke up and realized that their dreams had nasty surprises and hard questions waiting for an answer."

But should we be worried?

Well, it's complicated ;)

"

Life is a soup, and I am a fork.

"

- Anonymous.

Order your favorite Irish coffee and enjoy it with your loved ones. First things first. Then, let the show begin as the night is approaching ;)

PART
THREE

Living the new era

Technology has the potential to make us a better species, but at the same time, it poses a risk to our survival. How can we surf on the waves of technology without capsizing and drowning? I will provide you with some insights that will hopefully help you. You are the captain of your boat.

Tech lover that I am, I have always believed that technology has the potential to change the world for good. Its everlasting shine has brought wealth and convenience to the majority of us. No matter if we are rich or poor, if we come from a developing or a developed country, if we have a PhD or didn't complete high school.

I have described to you the main advantages and setbacks that we will face in the next 10 years, and I have insisted on the fact that for each new technological advance, there will be a reverse reaction that goes beyond the technology itself. If there were two things, and two things only, that you could remember from this book, they should be that, like it or not, every innovation has a broader influence than the innovation itself and that every discovery comes with a trade-off. We will have to judge if it is worth moving forward or not. And, let's be clear, moving forward is not always the best decision.

This last chapter of the book is my point of view about how to live in this changing world. This is an honest and open opinion based on scientific facts combined with a set of advice. No judgments here. This section goes beyond technology itself; it is more a guide to living well. I wrote this section because I realize that every individual's choices will decide the fate of the world and the place we leave to our kids when we die. Naturally, I want to help you make smart choices!

I believe there will be four main challenges in the years to come. Of course, there are more than four, but for me these are the most critical ones. They are:
- Climate change
- Privacy
- Education, work, and self-worth
- The world as a global village

A. Climate change

This first point has the potential to change the world as we know it. No surprises here, we are all aware of it. According to the UN's estimates, if our emissions of greenhouse gases don't change, the global mean temperature will rise 1-2°C by the end of the century.[318] The world's oceans will warm, and melting ice will raise the average sea level by 24–30 cm by 2065 and by 40–63 cm by 2100.[319] Even if emissions are stopped, the consequence of climate change will persist for many centuries, leading to irreversible changes in major ecosystems. To give a few examples, ecosystems as diverse as the Amazon rainforest and the Arctic tundra may be approaching thresholds of dramatic change through warming and drying. What's worse, mountain glaciers are in alarming retreat and the downstream effects of reduced water supply in the driest months will have repercussions that will transcend generations. Many islands may disappear from the maps, mass extinctions of species will be triggered, contamination in big cities will be inevitable, water pollution and drinking water scarcity will cause severe problems, and a long unpredictable fog will envelop all our dreams.[320]

While developed countries are trying to reduce their emissions, it is the turn of developing countries to have their industrial revolution with the *inevitable* environmental impact. There is no right or wrong answer to this dilemma, but I find that the only possible solution is to change our day-to-day habits to positively transform the world. It's going to be very hard, but we don't have a choice. I know that the following lines are not popular, but sooner or later, we will have to address them. Also, I believe that technology will play a role in monitoring our impact on the world. At a glance: move toward veganism, stay local, and think minimal. These three changes in our consumption habits are easier said than implemented. Movements are going in that direction, but I am also convinced that it has to go step by step.

---

[318] Taking as a reference the year 1990
[319] The reference scenario is the period of 1986-2005
[320] (United Nations, 2018)

To begin with, beef production and animal husbandry in agriculture is the number one cause of global warming and, therefore, of climate change. In other words, eating meat creates more pollution than driving a car. Methane (CH_4 emitted from livestock) and nitrous oxide (N_2O emitted during agricultural activities) have 32 and 280 times the Global Warming Potential of CO_2 respectively. It is a scientific fact,[321] [322] [323] [324] [325] even though the media, the governments, and the main environmental NGOs fail to treat this topic with the urgency it deserves.[326]

The population is growing, and we need to feed the 9.7 billion people, including a growing middle class that will be able to afford meat and dairy products, who will live on this planet in 2050.[327] If we want to maintain the earth as a livable place, we will have to move toward a more vegan way of living. This reality is very unpopular because most people don't want to change their dietary habits. Not even Al Gore dared to talk about it. But we have to take responsibility for the impact we are having on the world. Don't get me wrong. I was raised in a meat-eating family, and I love to eat a burger with extra bacon and cheese. But it simply isn't sustainable. I predict that this point will be of great controversy in the coming years and will be a source of conflict between the Y generation (millennials) and the Z generation. Meat consumption will have to be decreased, and a more vegan way of living may be the answer. I don't believe we need to go 100% plant-based, but what is clear is that our current diets are not sustainable.[328] [329]

---

[321] (Environmental Defense Fund, 2018)

[322] (United States Evironmental Protection Agency, 2018)

[323] (Rifkin, 2015)

[324] (United Nations, 2006)

[325] (IPCC, 2014)

[326] I have always been upset with the classic environmental NGOs like Greenpeace (with its 2.8 million members) that fight against climate change without daring to mention meat consumption. Hard questions are needed. Again, this shows the power of lobbies and how difficult it is to change people's habits.

[327] (United Nations, 2017)

[328] (Vox, 2017)

[329] (Rifkin, 2015)

The second point is to buy local. It is incredible how many miles our food flies to reach our homes. Let's remember that the second reason for climate change is transportation. The delicious avocado on your plate has probably traveled more than you will this year. It is sometimes very difficult to know the origin of certain products, but some cooperatives exist near your neighborhood that can supply you with local food. There are also more and more supermarkets that mention the origin of products clearly on their packages. Of course, balance is always key. Sweet papaya is a pleasure that you should have from time to time. But it is this consciousness of the impact that we have through our purchases that will make a difference in the middle term. If we want to reduce our transportation impact, buying local is an easy strategy to follow and might become one partial solution. In this sense, technology and the Internet of Things will improve the efficiency of the current transportation industry. At the regional and national level, it is important to increase the usage of public transportation and bikes among citizens.[330]

Last but not least, the concept of minimalism will need to enter our homes one day. The current pace of consumption, capitalism, and materialistic accumulation will not bring us the happiness or fulfillment we seek. There is a movement that says we should live with only 100 "things" and that this is enough. The point is not the number itself; the main lesson here is that we can live with less. Objects have a whole supply chain behind them that impacts the world in terms of waste, energy consumption, and raw material usage. Fast-moving consumer goods are destroying the earth, and the need for them is debatable. Americans are by far the most extreme example of consumerism. To have some numbers, the average American home has 300,000 items;[331] Research found that the average 10-year-old British kid owns 238 toys but plays with just 12 daily;[332] the average American woman today

---

[330] Regarding electric cars. Your car might be electric, but if that electricity is produced in a coal production plant, you are not solving the problem. ☺
[331] (Los Angeles Times, 2014)
[332] (Telegraph, 2010)

owns 30 outfits, while in 1930 that figure was 9;[333] Americans
spend $1.2 trillion annually on non-essential goods–in other
words, items they do not need[334]... the list just goes on. We can
do much better. If you see yourself in the examples and don't
know where to start, I recommend *The Life-changing Magic of
Tidying up* by the Japanese writer, Marie Kondo.

---

[333] (Forbes, 2015)
[334] (The Wall Street Journal, 2011)

B. Privacy

The second major challenge is, in my opinion, a completely overlooked topic. There hasn't been a serious global debate between world leaders about the level of encryption and anonymity that should be required for internet users and its service providers. Only when major digital leaks happen does the media raise a hue and cry about privacy. With the Internet of Things, everything and everyone will be connected, and the amount of data exchanged between "things" and "people" will increase exponentially. The risks coming from hacking and surveillance are still difficult to foresee, but in my opinion, privacy will become one of the most important challenges of this century. It is the responsibility of policymakers to set the correct rules for the game, the responsibility of users like us to understand the risks of the internet, the tech companies, and our behavior.

Here you have a list of advice that will protect you.[335] I know it is a long list for your busy agenda. I tried to keep it as short and non-geeky as I could. We also have to be aware that when you move to a more secure behavior, you will lose convenience. It is the price to pay, and unfortunately, there is no other way (yet). By following this advice, you are protecting your banking information, your medical data, your Facebook profile, your pictures, and your work, among many other things.

Let's begin:

- You need genuine software. There are a lot of security bugs that are often solved with updates.[336] If you have non-genuine software installed on your PC, you are

---

[335] I don't receive any commission for the references I mention in the following lines. They are simply software that I have installed myself, and I am happy with the results. Don't hesitate to look for other suppliers. The important thing is not who you choose but how protected you are.

[336] A secret between you and me: when I was younger, I used to have all the software of my computer hacked so I could play the latest games without paying. I was young and ignorant. I was not aware of all the risks that I was incurring. Don't make the same mistake I did.

putting your computer/smartphone at risk. There are a lot of free versions of almost any software (including operating systems). Check online for them. If for any reason you have to install non-genuine software, use a virtual machine (VirtualBox is an excellent open source option).

- You need protection on your laptops and smartphone. It is a must to have three things: a firewall, an updated anti-virus, and a VPN provider.[337] There are some free versions of these software, but I would go for the paid ones because they often offer interesting packages and routine updates. Just like you would spend money on a locker to protect your house, I would recommend paying for this protection. I would advise you to research and purchase the software that is most aligned with your needs and budget. I use Kaspersky Total Security as firewall and anti-virus, and NordVPN as my VPN provider.

- You need strong passwords. I know they are difficult to remember. I know you have lots of accounts. But we don't have a choice if we want protection. I recommend you have a password manager that will gather all your keys and encrypt them with a single master key. This allows you the luxury of having to remember just one key. I use the free and open source KeePass (available for MAC, Windows, iOS, and Android). And promise me that you won't use the same password for every online service. Another interesting tool that can be used with some providers is a two-step verification (first step with your laptop, second step with your smartphone). This is a great tool that can protect you (especially if you do mobile banking and/or exchange crypto-currencies).

---

[337] Simplifying a lot: firewalls prevent bad guys on the internet from accessing your network from their sofa; anti-virus avoids, detects, and removes malicious software; and a VPN allows your computer to create a secure connection over the internet.

- You need to change your digital habits to gain privacy and protect yourself. You shouldn't post anything online that could be used against you. The net doesn't forget. On the same point, always be careful with the USB sticks that you put in your computers and the links you click. Again, if you are not sure, it's always better to use a virtual machine.

- Tor is a web browser that allows you to swim on the net truly anonymously. Install it just for the day you will need it. It slows the traffic a bit, but it is the only browser that is truly anonymous. If you don't like Tor, use Brave or Firefox with at least the following free plug-ins installed: HTTPS Everywhere, uBlock Origin, NoScript, Stop Fingerprinting, and Windscrib. Three of them are free and open source.

- ProtonMail is a free and open source email service that allows you to send encrypted emails even if the receiver doesn't have an account with them. It has a simple and user-friendly interface, and you don't need any specific IT know-how to use it.

Of course, this list is not exhaustive and it has to be combined with good general practices. As a rule of thumb, don't share anything online that could compromise you in the future. You have to remember that the cloud is simply somebody else's computer.

Last but not least, your privacy IS important. And trust me, you definitely have lots of things to hide. And I am not only talking about the nude pictures you send to your partner.[338] This doesn't mean that you are doing something illegal or that you are a terrorist. The people you talk to, the jokes you make, the speed with which you reply, the topics you talk about, the attachments you send, your Google searches, the flights you book, the prices

---

[338] Funny statistic: 49% of internet users send or receive sexual content via video, photo, email, or messaging in a new concept called "sexting" (Scientific American, 2013). Somehow, this reality scares me.

you pay, etc., can give anyone a lot of information about who you are, what you believe in, how wealthy you are, how many friends you have, how trustworthy you can be, how many girlfriends (or boyfriends) you've had, your sexuality, your health, your purchasing power, your psychological stability, your next vote, etc. All this information must remain private and encrypted end-to-end. Always. The reality is that the world is still far from being secure. It is our duty to demand the highest standards of protection from our government, from our ISF (Internet Service Provider such as AT&T or Verizon), and from our other service providers (such as Google and Facebook). But it is also our duty to have the correct set of habits and attitude about our protection.[339] The reality is that data leaks will continue to happen, like it or not, even for tech giants. It is better to be ready for it.

---

[339] If you are a bit geeky and want to go the extra mile, I recommend watching the YouTube video *Becoming Anonymous: The Complete Guide To Maximum Security Online (2017/2018)* from Techlore.

C. Education, work, and self-worth

We have seen that technology will completely change the way we understand education, work, and our sense of self-worth. In the following lines, I will try to describe some advice for living in this new era.

First, fluent English is a must. Period. You need to know how to read, write, and speak if you want to survive in this global and competitive world. It is true that Chinese and Spanish have more native speakers than English, but the diversity of its speakers makes English the only truly global language. There are thousands of tools out there to learn English, from the classic classroom courses to fancy apps such as Duolingo, but I have always believed that you will only learn a language out of need and/or fun. Date one (or two) expats, travel alone to an unknown country and make not-from-your-country friends there, go to parties with internationals, watch Netflix in the original version, or check out John Oliver's comedy show. It is important that you have fun when you learn the language! If you don't, you will get tired very soon and lose motivation.

Second, please take advantage of technology and its global reach. Teamwork makes projects successful. If you don't have time or energy to do whatever can be done online, trust someone from the net and outsource your tasks. For example, I am not an English native speaker; this book has been written by me but proofread by two specialists who speak English much better than I ever will. I am not a graphic designer, so this cover and the interior design has been outsourced too. I am not a publisher, so the printing process is made by Amazon. Several people have participated in the creation of this book. And the interesting part is that they are all people I will probably never meet in person. Welcome to the world we are living in, and it's only going to get crazier.

On the same line, if you examine this book closely, you will find two words repeated over and over. The words are "change" and "technology." I truly believe in the power of change. Being flexible with the new technological changes will be the true added value of the leaders of tomorrow. I hope you will be more aware now that you have read this book. The reality is that "learning" is also a skill that you have to practice to master. Even if you have a stable job in a stable company in a stable country with a stable family, my advice would be to never stop learning! And by learning, I specifically mean studying. And the cool point is that learning is free! Learn whatever you like—how to hack your friend's computer, how to play an instrument, how to use an augmented reality tool, or how to make a perfect gin and tonic. You never know what lessons will come handy in the future, but you know that it will require the skill "learning." Remember, like it or not, it is very likely that your current job will disappear in the future.

Last, but for me the most important: travel and mingle with people who are different from you. In this global city that we are entering, the value of the human workforce is in understanding how the globalized world works and what the interconnections between its different players are. Understand the local working culture and its unique needs. In the old times, trade and mindset were national. Now they have become global. It will be your job to listen to what other cultures have to teach you. Don't be afraid!

D. The world as a global village

My last point is connected with governments and global organizations. In my opinion, they will have three enormous challenges to overcome. In general, it is important to always keep in mind that the one thing governments cannot do is go against technology. Change and innovation will always be there; so don't try to go against technology. Don't try to go against automation to save jobs. It has never worked, and it never will. Instead, I would advise embracing the coming changes and regulating them quickly according to the ethics and the reality of the country itself. It won't be easy. But I believe that setting legal frameworks for new technologies is the key to succeeding in our technological revolution and promoting healthy innovation.

First, it is important to develop a culture of change and innovation. I already know what you are thinking: this is exactly what every politician says but never does. I know that culture is always difficult to change and new legislation will take time to give results, but I believe that governments have a role with three main axes:

Updated education: We are not in the industrial era anymore. We are entering a new paradigm that will need a new set of skills. To begin with, subjects about entrepreneurship, creativity, and innovation are a must. Everything related to standardized tests should be removed because standardized tests don't reflect what the kids need to learn. In the old times, standardized tests made sense because the workforce was required to follow processes and apply them in factories (which has a lot in common with the process of filling a standardized test). In the new reality, workers are expected to be creative, work in teams, and be good communicators. You will never be able to monitor those skills with fixed exams.

Collaboration schools/university-companies: Let's be clear. Whatever your political ideas are, in the capitalistic economies (i.e., almost all countries in the world) the private sector sustains the whole economy, including the public sector. Schools need more collaboration programs with companies and business leaders. The same applies in reverse. Like it or not, private companies are the ones that will always have a say in the future of the country. Kids should visit farms, factories, even offices to know more about the reality of work and how it is changing.

Quicker legislation: Traditionally (and especially in Europe), legislation is always late. We need more policymakers who are willing to move faster. Just to give an example, bitcoins were created in 2009, and we still don't have clear legislation about how to use them. Wake up, governments! We need fast legislation to set the correct frameworks for innovation to happen and develop.

The second point is about global governance. We are living in a world with an unpredictable future, and globalization has made interdependencies between countries the new normal. The challenge is that it has also brought global problems that can only be solved with global solutions. Countries will, *unfortunately*, have to talk and make concessions if we want to survive. Climate change, privacy, refugee crises, tax havens, terrorism, international espionage, immigration controls, international conflicts, disease prevention—all require countries working together to make decisions. Push your legislators and lobby groups toward better global governance. Concessions in sovereignty are the price to pay to get out of the global mess we are all in.

My third point is regarding cities. If you are looking for development, you must move to one of them. In a city, you will find the opportunities and the talent you are looking for. In 2016, 45 cities had 5-10 million inhabitants. By 2030, 10 of these are projected to become megacities (cities with more than 10 million residents). Projections indicate that 29 additional cities will cross the 5 million mark between 2016 and 2030. By 2030, 60% of the world population will live in cities.[340] It is there that you will meet the people who will help you achieve your dreams.

Last but not least, I believe that while keeping a global mindset is important, at the end of the day, we live in small communities even if we are in crowded cities. Try to think global while acting local. Be involved in local organizations that you know and believe in. Help your neighbors when their boiler breaks. Support a friend who is trying to start a business. Smile and have fun. Life is short, so always stay humble even if you become successful and remember to remain loyal to your family, friends, and values.

---

[340] (United Nations, 2016)

In this chapter, I have tried to give you some of the tips that I believe will help you navigate through the years to come. It is our responsibility to leave a better world for our kids. Our fate will be defined by the individual decisions each one of us makes. If we all influence our network positively using technology as a tool for good, we will make this world a better place for the future generations. Isn't life exciting?

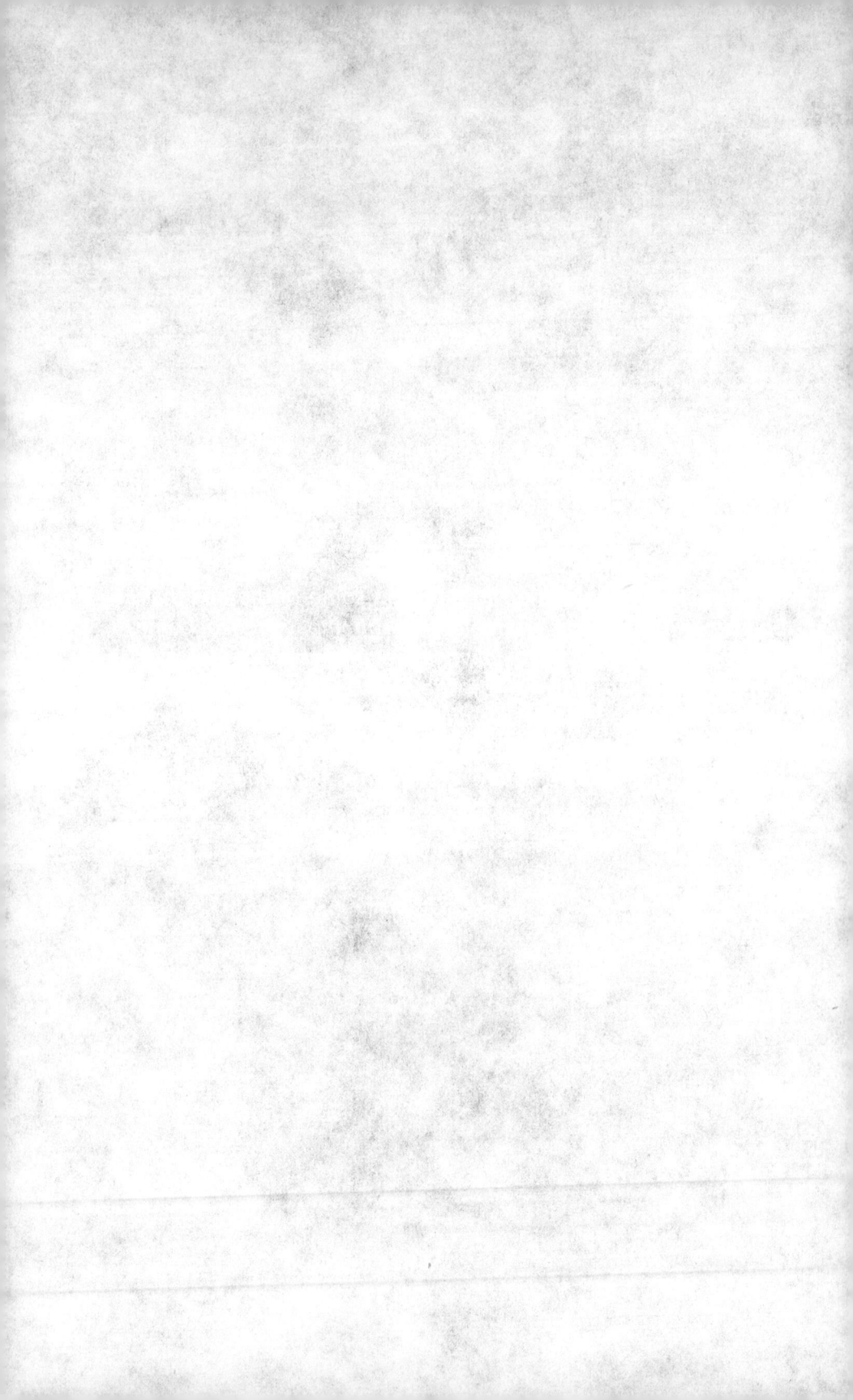

HAPPY END?

I want to end this book with an observation. Humankind has lived through several huge technological transformations. From the discovery of fire to the first steam machine, we have always survived and been better off after the revolution. The world today is much better than it was before. Much, much better! Absolute poverty has decreased from 57% to under 17% over the last 30 years; mortality rates for children under five have dropped by 50% in the last 25 years; global literacy rates have increased from around 10% to close to 100% in the last 500 years; the number of children in hazardous work conditions and performing child labor has been reduced by more than 50% in the last 16 years...[341] There are reasons to be hopeful. Indeed, there have always been winners and losers in any technological revolution. But this shouldn't be scary. Adaptation is what makes humans a great species. I believe that we are in an exciting new phase for human development. If we treat technology with ingenuity and without forgetting our critical thinking, nobody will be able to stop us :)

How do we want to be remembered?

---

[341] (SingularityHub, 2016)

ACKNOWLEDGMENTS

This book couldn't have become a reality without the support and love of the people described in the following lines.

If you liked the book, it is simply because I am lucky enough to have incredible people around who supported me during the whole torturous process of writing.

If you didn't like the book, it is simply because I am not yet good enough for you. I will keep on trying ;).

Primero de todo, **muchas gracias** a mis padres. Nunca hubiese llegado donde estoy ahora si no fuese por vosotros. Sois mis héroes.

Muchas gracias a mis hermanas María y Helena. Sois con diferencia el mejor regalo que nunca he recibido.

Muchas gracias a mi familia. A mis titas Conchita y María José, a mi tito Juan Ramón y a mis primas Paloma y María José por todos los momentos en familia que hemos pasado.

Thank you to Maria Varela for all the love and care. When I look at you, I wonder how I got to be so lucky. You are like a dream come true.

Thank you to my Fundamentals (Lorenzo Brizzi, Lorin Sleeuwagen, Arturo Ambrosio, Yannick Hippolyte, Amir Jouadi, and Zakaria Sellam) for being the exact definition of what true brotherhood means.

Thank you to Marta González for being that loyal friend that everyone should have and admire.

Thank you to Jean-Luc Verhelst for your passion and entrepreneurial spirit that brought us where we are.

Thank you to my Belgian crew (Javier Lázaro, Arantxa Múgica, Elena Ramírez, Sinouhe Monteiro, Mar Jorba, Amandine Capras, Juan Luis Rousselot, Ruth Bernabé, Sara Cabrejas, Dalil Djinnit, Marianna Matichecchia, Saad Shuja, Gabriela Fariña, Zincke Vandenbroucke,

Suzanne Safi, Stefano Brizzi, Maitane Serna, Manuel Ortegon, Carmen Martínez, Joaquin Rivas, Jacopo Troise) for all the love and smiles that we shared.

Thank you to my international buddies (Marina Sánchez, Nicole Mezgolits, Begoña Pla, Shamma Raghib, Sandra Herrera, Cristina Rucareanu, Ioanna Kordali, Julia Sadykova, Vinthujah Bala, Marlén Villena, Orsolya Katalin) for all the laughs and wine we had together.

Thank you to my Whales core team (Alberto Moral, Victor Lucena, and Robert Garrigosa) for reminding me that funny jokes need to be shared.

Thank you to my Barcelona crew (Maximiliano Fernández, Marià Torrecilla, Guillem Orpinell, Bernat de Miguel, Cristina Massons, Cristina Benito, Albert Garcia, Jenifer González, Albert Poti, Dani Rey, Jordi García, Ania Alay, Marta Saumell, Mireia Penya, Daniel Peña, Catu Máiz, Jessica Rubio, Pau Jove, and Andrea Marchan) for simply being awesome.

Thank you to my Siemens team (special mention to Marc Renard, Rachid Boujabed, Thierry Ndach, Jammale Lamarti, Daisy Plantefeve, Caroline Cottyn, Mohamed Lemine, Maria-Paz Martens, Arthur Charlier, and to all the Energy Department) for all the good moments.

Thank you to the Board of Europeans Students of Technology for everything I learned there as a student.

Thank you to all the companies that are trying to make the internet a safe, open, and transparent space accessible to all. I am especially inspired by the Open Source Initiative, Mozilla, Creative Commons, GNU, and Wikipedia.

Thank you to Edward Snowden for showing the world the power of technology and the importance of privacy.

Thank you to the stoicism philosophy for being an inspiration in my day-to-day life.

BIBLIOGRAPHY

Alfaro, L., & Maggie, C. (2014). The Global Agglomeration of Multinational Firms. Institute for
International Economic Policy. Washington: George Washington University. Retrieved from
https://www2.gwu.edu/~iiep/assets/docs/papers/2014WP/ChenIIEPWP201420.pdf

Ankerl, G. (2000). Coexisting Contemporary Civilizations: Arabo-Muslim, Bharati, Chinese and Western.
INU Press.

AquaBounty. (2017). AquaBounty believes that aquaculture can - and must - do better. Retrieved January
2017, from https://aquabounty.com/sustainable/

Atlantic Council. (2016). Smart Homes and the Internet of Things. Retrieved from
https://otalliance.org/system/files/files/initiative/documents/smart_homes_0317_web.pdf

Automation.com. (2014, February 19). The 4th Industrial Revolution, Industry 4.0, Unfolding at Hannover
Messe 2014. Retrieved from https://www.automation.com/automation-news/article/the-4th-
industrial-revolution-industry-40-unfolding-at-hannover-messe-2014

Baoying Wang, R. L. (2015). Big Data Analytics in Bioinformatics and Healthcare. Hershey: Medical
information science reference.

BBC. (2013, July 1). Q&A: Reform of EU farm policy. BBC. Retrieved from
http://www.bbc.com/news/world-europe-11216061

BBC. (2015, September 11). Will a robot take your job? Retrieved from Technology site:
http://www.bbc.com/news/technology-34066941

BBC. (2016, June 12). Here's the truth about planned obsolescence of tech. Retrieved from
http://www.bbc.com/future/story/20160612-heres-the-truth-about-the-planned-
obsolescence-of-tech

BBC. (2016, June 16). Here's the truth about the ·planned obsolescence· of tech. Retrieved from
http://www.bbc.com/future/story/20160612-heres-the-truth-about-the-planned-
obsolescence-of-tech

BBC. (2018, January 8). Apple investigated by France for 'planned obsolescence'. Retrieved from
http://www.bbc.com/news/world-europe-42615378

BBVA. (2016). Artificial Intelligence is driving the definitive automation of financial services. Fintech &
Innovation. Retrieved from https://www.bbva.com/en/news/science-
technology/technologies/artificial-intelligence-driving-definitive-automation-financial-
services/

Berg Insight. (2015). Smart Homes and Home Automation. Retrieved from
https://ec.europa.eu/research/innovation-union/pdf/active-healthy-
ageing/berg_smart_homes.pdf

Bershidsky, L. (2014, December 30). A great year for the far right. The Japan Times. Retrieved from
http://www.japantimes.co.jp/opinion/2014/12/30/commentary/world-commentary/great-
year-far-right/#.WKTDtG8rKCg

Big Thinking. (2016). Jeremy Rifkin on the Energy Internet. (B. T. editors, Ed.)

Bolívar, M. P. (2017). Smart Technologies for Smart Governments: Transparency, Efficiency and
Organizational Issues (Public Administration and Information Technology). Springer.

Brown, R. G. (2013, November 24). https://gendal.me. Retrieved December 13, 2016, from
https://gendal.me/2013/11/24/a-simple-explanation-of-how-money-moves-around-the-
banking-system/

Business Insider. (2016, July 20). Here's what it's like to work in a WeWork building, the $16 billion
company that simulates startup life. Retrieved from http://www.businessinsider.com/working-
in-a-wework-2016-7?international=true&r=US&IR=T

Business Insider. (2017). Online ad revenues are surging, but 2 companies are getting most of the spoils.
Business Insider. Retrieved from http://www.businessinsider.com/online-ads-revenues-going-
to-google-and-facebook-the-most-2017-4?IR=T

Canadian Government. (2010, January). Cities and Growth: Earnings Levels Across Urban and Rural Areas: The Role of Human Capital. Retrieved from http://www.statcan.gc.ca/pub/11-622-m/2010020/part-partie1-eng.htm#h2_7

Carlson, R. (2017). Megatech: Technology in 2050 (Vol. Biotechnology's possibilities). London: The Economist.

Climate Central. (2013, May 3). The Last Time CO2 Was This High, Humans Didn't Exist. Retrieved from http://www.climatecentral.org/news/the-last-time-co2-was-this-high-humans-didnt-exist-15938

CNN. (2008). Designer babies: Creating the perfect child. Retrieved from http://edition.cnn.com/2008/TECH/science/10/30/designer.babies/index.html

CNN. (2016). Healing the future. Retrieved from CNN international Edition: http://edition.cnn.com/interactive/2014/04/health/the-cnn-10-healing-the-future/

Coindesk. (2018, April 19). $6.3 Billion: 2018 ICO Funding Has Passed 2017's Total. Retrieved from https://www.coindesk.com/6-3-billion-2018-ico-funding-already-outpaced-2017/

Connolly, K. (2013, January 9). Mass donor organ fraud shakes Germany. The Guardian. Retrieved from https://www.theguardian.com/world/2013/jan/09/mass-donor-organ-fraud-germany

Da Vinci Surgery. (2017). Retrieved from da Vinci® Surgery: Minimally Invasive Surgery: http://www.davincisurgery.com/

Data&Society Research Institute. (2016). Nonconsensual image sharing: one in 25 Americans has been a victim of "Revenge Porn". New York: Center for Innovative Public Health Research. Retrieved from https://datasociety.net/pubs/oh/Nonconsensual_Image_Sharing_2016.pdf

Deepwater. (2010, April 26). Transocean: deepwater horizon drills world's deepest oil&gas well. Retrieved from https://web.archive.org/web/20100426171257/http://www.deepwater.com/fw/main/IDeepwater-Horizon-i-Drills-Worlds-Deepest-Oil-and-Gas-Well-419C151.html

Deloitte. (2015). Challenges and solutions for the digital transformation and use of exponential technologies. Zurich: The Creative Studio at Deloitte. Retrieved from https://www2.deloitte.com/tw/en/pages/manufacturing/articles/industry4-0.html

Deloitte. (2015). Disrupting the Financial Services Industry? Deloitte Touche Tohmatsu Limited. Retrieved from https://www2.deloitte.com/content/dam/Deloitte/ie/Documents/FinancialServices/IE_Cons_Blockchain_1015.pdf

Deutschland. (2014, April 2). Industry 4.0 at Hannover Messe. Retrieved from https://www.deutschland.de/en/topic/business/globalization-world-trade/industry-40-at-hannover-messe

Digital Trends. (2016, September 30). Retrieved from High electricity bill? Four ways to measure home energy consumption: http://www.digitaltrends.com/home/how-to-measure-home-energy-use/

Donovan, T. (2011). State Of The Ocean: 'Shocking' Report Warns Of Mass Extinction From Current Rate Of Marine Distress. The Huffington Post. Retrieved from http://www.huffingtonpost.com/2011/06/20/ipso-2011-ocean-report-mass-extinction_n_880656.html

Dow. (2015, December 11). Retrieved from Dupont and Dow to combine in merger of equals: http://www.dow.com/en-us/news/press-releases/dupont-and-dow-to-combine-in-merger-of-equals

Drift for transition. (2018). PROSEU: Active participation of citizens into Europe's transition towards fair and clean energy. Retrieved from https://drift.eur.nl/projects/proseu-renewable-energy-prosumerism/

El País. (2017, April 16). Los últimos mineros mileuristas. Retrieved from https://elpais.com/economia/2017/04/11/actualidad/1491909951_969594.html

Environmental Defense Fund. (2018). Methane: The other important greenhouse gas. Retrieved from https://www.edf.org/methane-other-important-greenhouse-gas

EOS. (2018). Retrieved from Aerospace: Vectoflow - Additive Manufacturing of probes for measuring speed and temperature in turbo engines: https://www.eos.info/aerospace-vectoflow-additive-manufacturing-of-probes-for-measuring-speed-and-temperature-in-turbo-engines-ea0691d8a20ee1eb

Eric D. Williams, R. U. (2002, October 25). The 1.7 Kilogram Microchip: Energy and Material Use in the Production of Semiconductor Devices. Retrieved from http://pubs.acs.org/doi/abs/10.1021/es025643o

European Comission. (2014). Subsidies and costs of EU energy. Directorate-General for Energy. European Comission. Retrieved from https://ec.europa.eu/energy/sites/ener/files/documents/ECOFYS%202014%20Subsidies%20and%20costs%20of%20EU%20energy_11_Nov.pdf

European Comission. (2016, September 6). About TTIP - basics, benefits, concerns. Retrieved from http://ec.europa.eu/trade/policy/in-focus/ttip/about-ttip/questions-and-answers/

European Parliament. (2014). An overview of Europe's film industry. Members' Research Service. Retrieved from http://www.europarl.europa.eu/RegData/etudes/BRIE/2014/545705/EPRS_BRI(2014)545705_REV1_EN.pdf

Ewala.co. (2018). Ewala. Retrieved from http://www.ewala.co

Fairphone. (2018, February 20). The modular phone that's built to last. Retrieved from https://www.fairphone.com/en/

Fast Company. (2013, March 22). The Concept Of Wabi-Sabi, And Why Perfection Is The Wrong Goal. Retrieved from https://www.fastcompany.com/3007322/concept-wabi-sabi-and-why-perfection-wrong-goal

FireflySixtySeven. (2014, November 2). Own work using Inkscape, based on Maslow's paper, A Theory of Human Motivation. (Creative Commons Attribution-Share Alike 4.0 International license). Retrieved from https://commons.wikimedia.org/wiki/File:MaslowsHierarchyOfNeeds.svg

Forbes. (2011, November 19). Peggy Noonan On Steve Jobs And Why Big Companies Die. Retrieved from http://www.forbes.com/sites/stevedenning/2011/11/19/peggy-noonan-on-steve-jobs-and-why-big-companies-die/#45a414803e57

Forbes. (2012, April 27). A Glimpse At A Workplace Of The Future: Valve. Retrieved from https://www.forbes.com/sites/stevedenning/2012/04/27/a-glimpse-at-a-workplace-of-the-future-valve/#d51f12475577

Forbes. (2013, November 26). Growing The Use Of Drones In Agriculture. Retrieved from https://www.forbes.com/sites/rakeshsharma/2013/11/26/growing-the-use-of-drones-in-agriculture/#60899d943ea7

Forbes. (2015, December 28). Kensho: The Financial Answer Machine. Retrieved from https://www.forbes.com/sites/laurashin/2015/12/09/kensho-the-financial-answer-machine/2/#128c8f9539ca

Forbes. (2015, July 22). The Complete Guide To The 5 Types Of Organizational Structures For The Future Of Work. Forbes. Retrieved from https://www.forbes.com/sites/jacobmorgan/2015/07/22/the-complete-guide-5-types-of-organizational-structures-for-the-future-of-work/#143a701c7705

Forbes. (2015, January 15). The real cost of shopping habits. Retrieved from https://www.forbes.com/sites/emmajohnson/2015/01/15/the-real-cost-of-your-shopping-habits/

Forbes. (2016, January 31). Is Social Media The Future Of Healthcare? Retrieved from Advisor Network: http://www.forbes.com/sites/joannabelbey/2016/01/31/is-social-media-the-future-of-healthcare/#60f899db48d6

Forbes. (2016, October 26). The Future Of Health Care Is In Data Analytics. Retrieved from Entrepreneurs /Big Data: http://www.forbes.com/sites/mikemontgomery/2016/10/26/the-future-of-health-care-is-in-data-analytics/#3ba6b8ad2f71

Fortune. (2015, October 12). What San Francisco taught BMW Group about car-sharing. Retrieved from http://fortune.com/2015/10/12/san-francisco-bmw-car-sharing/

Fortune. (2016, June 22). Chronic Loneliness Is a Modern-Day Epidemic. Fortune. Retrieved from
http://fortune.com/2016/06/22/loneliness-is-a-modern-day-epidemic/

Frey, C. B., & Osborne, M. A. (2013). The future of employment: how susceptible are jobs to
computerisation? Oxford: Oxford University. Retrieved from
http://www.oxfordmartin.ox.ac.uk/downloads/academic/The_Future_of_Employment.pdf

Future Human Evolution. (2017). Genetic Engineering Human Cloning. Retrieved from
http://futurehumanevolution.com/genetic_engineering_human_cloning

Future Today Institute. (2017). Tech trend report. New York: Future Today Institute.

Futurity. (2018, April 25). New A.I. application can write its own code. Retrieved from Science and
Technology: https://www.futurity.org/artificial-intelligence-bayou-coding-1740702/

Gallup. (2016, May 12). Business Journal. Retrieved from Millennials: The Job-Hopping Generation:
http://news.gallup.com/businessjournal/191459/millennials-job-hopping-generation.aspx

Gartner. (2015, December 17). The Future of the Energy Grid. Retrieved December 20, 2016, from
http://www.gartner.com/smarterwithgartner/the-future-of-the-energy-grid/

Gartner. (2015, November 10). Gartner Says 6.4 Billion Connected "Things" Will Be in Use in 2016, Up 30
Percent From 2015. Retrieved from http://www.gartner.com/newsroom/id/3165317

Gartner. (2017, February 7). Gartner Says 8.4 Billion Connected "Things" Will Be in Use in 2017, Up 31
Percent From 2016. Retrieved from https://www.gartner.com/newsroom/id/3598917

General Electric. (2015). Annual report 2015.

General Electric. (2018, March). Predix Platform: The Foundation for Digital Industrial Applications.
Retrieved from https://www.ge.com/digital/predix-platform-foundation-digital-industrial-
applications

Genomes 2 People. (2018). The BabySeq Project. Retrieved from
http://www.genomes2people.org/babyseqproject/

Gentle Labs. (2017). Retrieved from About section: https://www.gentlelabs.com/about

Ginger.io. (2017). Retrieved from Home page: https://ginger.io/

Gizmodo. (2018, January 5). Artificial Genome Scientists Want to Build Human Cells That Are Impervious
to Viruses. Retrieved from https://gizmodo.com/artificial-genome-scientists-want-to-build-
human-cells-1825693446

Goldman Sachs. (2016). Virtual and Augmented Reality. The Goldman Sachs Group. Retrieved from
http://www.goldmansachs.com/our-thinking/pages/technology-driving-innovation-
folder/virtual-and-augmented-reality/report.pdf

Google. (2016, June 20). I'm feeling Yucky :(Searching for symptoms on Google. Retrieved from Google
Blog: https://blog.google/products/search/im-feeling-yucky-searching-for-symptoms/

Green Palm Sustainability. (2017). What is palm oil used for? Retrieved from http://greenpalm.org/about-
palm-oil/what-is-palm-oil/what-is-palm-oil-used-for

Greenpeace. (2016, March 3). Palm oil: who's still trashing forests? Retrieved from Making Waves:
http://www.greenpeace.org/international/en/news/Blogs/makingwaves/palm-oil-whos-still-
trashing-forests/blog/55724/

Hackernoon. (2017, May 9). How Much Time Do People Spend on Their Mobile Phones in 2017?
Retrieved from https://hackernoon.com/how-much-time-do-people-spend-on-their-mobile-
phones-in-2017-e5f90a0b10a6

Harari, Y. N. (2015). Sapiens: A Brief History of Humankind. Harper.

Huffington Post. (2014, December 18). We Already Grow Enough Food For 10 Billion People – and Still
Can't End Hunger. Retrieved from https://www.huffingtonpost.com/eric-holt-gimenez/world-
hunger_b_1463429.html

Hummod. (2017, January 15). Retrieved from http://hummod.org/

IBM. (2016, December 17). Delivering New Models Of Care With Cognitive Technology. Retrieved from
Watson Health Perspectives: https://www.ibm.com/blogs/watson-health/delivering-new-
models-care-cognitive-technology/

IBM. (2017). Predictive Analytics in Value-Based Healthcare: Forecasting Risk, Utilization, and Outcomes.
New York: IBM Corporation. Retrieved 2017, from https://www-01.ibm.com/common/ssi/cgi-
bin/ssialias?htmlfid=HLW03032USEN&

IBM. (2018, March 3). What is Big Data Analytics. Retrieved from
https://www.ibm.com/analytics/hadoop/big-data-analytics

IHS Markit. (2017, August 14). Global Smart Home Market to Exceed $14 Billion in 2017. Retrieved from
https://technology.ihs.com/594650/global-smart-home-market-to-exceed-14-billion-in-2017

IK4Tekniker. (2016, March 7). Industry 4.0: the computerization of manufacturing. Retrieved from
http://www.tekniker.es/en/industry-4-0-the-computerization-of-manufacturing

Independent. (2014, October 7). There are officially more mobile devices than people in the world.
Retrieved from http://www.independent.co.uk/life-style/gadgets-and-tech/news/there-are-
officially-more-mobile-devices-than-people-in-the-world-9780518.html

Independent. (2016, March 11). Narcissism: The science behind the rise of a modern 'epidemic'.
Retrieved from http://www.independent.co.uk/news/science/narcissism-the-science-behind-
the-rise-of-a-modern-epidemic-a6925606.html

Independent. (2017, July 31). Facebook's artificial intelligence robots shut down after they start talking to
each other in their own language. Retrieved from http://www.independent.co.uk/life-
style/gadgets-and-tech/news/facebook-artificial-intelligence-ai-chatbot-new-language-
research-openai-google-a7869706.html

Inés Figueiredo. (2017, December). Conversations in Cambodia.

International Energy Agency. (2016). Key world Energy Statistics. Paris: International Energy Agency.

International Energy Agency. (2017). Key world energy statistics. Retrieved from
https://www.iea.org/publications/freepublications/publication/KeyWorld2017.pdf

International Monetary Fund. (2017, October 24). Retrieved from World Economic Outlook Database:
http://www.imf.org/external/pubs/ft/weo/2017/02/weodata

International Monetary Fund. (2017, February 19). About the IMF. Retrieved from
http://www.imf.org/external/about.htm

International Programme on the State of the Ocean. (2013). The State of the Ocean report 2013.
Retrieved from http://www.stateoftheocean.org/wp-content/uploads/2015/10/State-of-the-
Ocean-2013-report.pdf

International Telecommunication Union. (2017). Measuring the information society report. Geneva,
Switzerland. Retrieved from https://www.itu.int/en/ITU-
D/Statistics/Documents/publications/mis2014/MIS2014_without_Annex_4.pdf

Investing News. (2018, May 16). 10 Top Cybersecurity Companies. Retrieved from
https://investingnews.com/daily/tech-investing/cybersecurity-investing/top-cyber-security-
companies/

Investing.com. (2018, August 20). Crypto-currencies. Retrieved from
https://www.investing.com/crypto/currencies

Investopedia. (2017, November 28). Top 3 Crowdfunding Platforms of 2017. Retrieved from
https://www.investopedia.com/small-business/top-crowdfunding-platforms/

IOT for all. (2017, December 4). 3 Examples of Successful Marketing with Virtual Reality. Retrieved from
https://www.iotforall.com/examples-of-marketing-with-virtual-reality/

IPCC. (2014). Climate Change 2014: Impacts, Adaptation, and Vulnerability. Summaries, Frequently
Asked Questions, and Cross-Chapter Boxes. Retrieved from
https://www.ipcc.ch/pdf/assessment-report/ar5/wg2/WGIIAR5-
IntegrationBrochure_FINAL.pdf

Jeffery, S. (2003, June 23). The EU common agricultural policy. The Guardian. Retrieved from
https://www.theguardian.com/world/2003/jun/26/eu.politics1

Jonas, H. (1979). The Imperative of Responsibility: In Search of Ethics for the Technological Age.

Kaser, J. (2016, February 09). Siemens CEO Joe Kaeser on the Next Industrial Revolution. Strategy
Business. (D. Gross, Interviewer)

Khokale, S. (2017, October 24). 4 Ways Wearables Are Changing the Future of Healthcare. Retrieved
from https://www.einfochips.com/blog/4-ways-wearables-are-changing-the-future-of-
healthcare/#readmore

Kondo, M. (2014). The Life-Changing Magic of Tidying Up: The Japanese Art of Decluttering and
Organizing. Ten Speed Press.

Los Angeles Times. (2014, March 21). For many people, gathering possessions is just the stuff of life. Retrieved from http://articles.latimes.com/2014/mar/21/health/la-he-keeping-stuff-20140322

Luppicini, R. (2010). Technoethics and the evolving knowledge society. IGI Global.

MaidSafe. (2017, January 16). A secure home for all your data. Retrieved from https://maidsafe.net/

Marous, J. (2017). The Use of AI in Banking is Set to Explode. The financial brand. Retrieved January 17, 2017, from https://thefinancialbrand.com/63322/artificial-intelligence-ai-banking-big-data/

McKinsey & Company. (2003). New horizons: Multinational company investment in developing economies. McKinsey Global Institute. Retrieved from http://www.mckinsey.com/business-functions/digital-mckinsey/our-insights/new-horizons-for-multinational-company-investment

McKinsey & Company. (2007). Defusing the pension time bomb. London: McKinsey & Company Media. Retrieved from http://www.mckinsey.com/~/media/mckinsey/dotcom/client_service/Public%20Sector/PDFS/McK%20on%20Govt/Inaugural%20edition/TG_defusing_the_pension_time_bomb.aspx

McKinsey & Company. (2012). The global company's challenge. McKinsey Quaterly. Retrieved from https://www.mckinsey.com/business-functions/organization/our-insights/the-global-companys-challenge

McKinsey & Company. (2013, April). The big-data revolution in US health care: Accelerating value and innovation. Retrieved from http://www.mckinsey.com/industries/healthcare-systems-and-services/our-insights/the-big-data-revolution-in-us-health-care

McKinsey & Company. (2014). 3-D printing takes shape. McKinsey Quarterly. Retrieved from http://www.mckinsey.com/business-functions/operations/our-insights/3-d-printing-takes-shape

McKinsey & Company. (2014, July). Healthcare's digital future. Retrieved from https://www.mckinsey.com/industries/healthcare-systems-and-services/our-insights/healthcares-digital-future

McKinsey & Company. (2015). Agility: It rhymes with stability. McKinsey Quarterly. Retrieved from https://www.mckinsey.com/business-functions/organization/our-insights/agility-it-rhymes-with-stability

McKinsey & Company. (2016). Bracing for seven critical changes as fintech matures. Financial Services. Retrieved from https://www.mckinsey.com/industries/financial-services/our-insights/bracing-for-seven-critical-changes-as-fintech-matures

McKinsey & Company. (2016). Energy 2050: Insights from the ground up. Oil & Gas insights. Retrieved from http://www.mckinsey.com/industries/oil-and-gas/our-insights/energy-2050-insights-from-the-ground-up

McKinsey & Company. (2016). Exploring the disruptive potential of synthetic biology. Interview, Pharmaceuticals & Medical Products. Retrieved from http://www.mckinsey.com/industries/pharmaceuticals-and-medical-products/our-insights/exploring-the-disruptive-potential-of-synthetic-biology

McKinsey & Company. (2016). How blockchains could change the world. High Tech. Retrieved from http://www.mckinsey.com/industries/high-tech/our-insights/how-blockchains-could-change-the-world

McKinsey & Company. (2016). Industry 4.0 after the initial hype: Where manufacturers are finding value and how they can best capture it. McKinsey Digital. Retrieved from https://www.mckinsey.de/files/mckinsey_industry_40_2016.pdf

McKinsey & Company. (2016). Renewable energy: Evolution, not revolution. Oil & Gas insights. Retrieved from http://www.mckinsey.com/industries/oil-and-gas/our-insights/renewable-energy-evolution-not-revolution

McKinsey & Company. (2016). Where machines could replace humans—and where they can't (yet). McKinsey Quarterly. Retrieved from http://www.mckinsey.com/business-functions/digital-mckinsey/our-insights/where-machines-could-replace-humans-and-where-they-cant-yet?cid=other-eml-ttn-mkq-mck-oth-1612

McKinsey & Company. (2017). Harnessing automation for a future that works. McKinsey Global Institute. Retrieved from http://www.mckinsey.com/global-themes/digital-disruption/harnessing-automation-for-a-future-that-works?cid=other-eml-alt-mgi-mgi-oth-1701

McKinsey & Company. (2017). Jobs lost, jobs gained: workforce transitions in a time of automation. McKinsey Global Institute. Retrieved from https://www.mckinsey.com/~/media/McKinsey/Global%20Themes/Future%20of%20Organiz ations/What%20the%20future%20of%20work%20will%20mean%20for%20jobs%20skills%20 and%20wages/MGI-Jobs-Lost-Jobs-Gained-Report-December-6-2017.ashx

McKinsey & Company. (2017). Security in the Internet of Things. McKinsey Article. Retrieved from http://www.mckinsey.com/industries/semiconductors/our-insights/security-in-the-internet-of-things

McKinsey & Company. (2108, March). How artificial intelligence and data add value to businesses. San Francisco, United States. Retrieved from https://www.mckinsey.com/global-themes/artificial-intelligence/how-artificial-intelligence-and-data-add-value-to-businesses?cid=other-eml-alt-mgi-mgi-oth-1804&hlkid=5c3e7b9384bd4992998d2a188dc2ae76&hctky=9884620&hdpid=6ba01c21-bdc1-4c53-87d2-b5573121ef7f

Medium. (2016, January 6). Why privacy is important, and having "nothing to hide" is irrelevant. Retrieved from https://medium.com/@obsidian_crypto/why-privacy-is-important-and-having-nothing-to-hide-is-irrelevant-d011b49de4c8

Merrill Lynch. (2017). Finances in Retirement: New Challenges, New Solutions. Bank of America Corporation. Retrieved from https://www.ml.com/articles/age-wave-survey.html

Mirmiki. (2018). Suport tècnic per a particulars, venim i t'ajudem. Retrieved from https://www.mirmiki.com/

Mona Hammami, R. A. (2016). Looking Ahead: The 50 Trends That Matter. Xlibris.

Morgan, J. (2014). The Future of Work: Attract New Talent, Build Better Leaders, and Create a Competitive Organization. Wiley; 1 edition (August 25, 2014).

Nasdaq. (2017, August 10). What is an ICO? Retrieved from https://www.nasdaq.com/article/what-is-an-ico-cm830484

National Health Service UK. (2015, March 13). Loneliness 'increases risk of premature death'. Retrieved from https://www.nhs.uk/news/mental-health/loneliness-increases-risk-of-premature-death/

National Institutes of Health. (2016). World's older population grows dramatically. Retrieved from https://www.nih.gov/news-events/news-releases/worlds-older-population-grows-dramatically

Nature. (2012, May 10). Comparing the yields of organic and conventional agriculture. Retrieved from https://www.nature.com/articles/nature11069

New Zeland Herald. (2018, February 3). School tells parents: Ban your kids from Facebook. Retrieved from http://www.nzherald.co.nz/nz/news/article.cfm?c_id=1&objectid=11986267

Nike. (2017, January 10). Retrieved from Nike Plus: http://www.nike.com/be/en_gb/c/nike-plus

Obama, B. (2013). You can't have 100% security and 100% privacy' Obama defends NSA's secret 'data mining. Retrieved from https://www.youtube.com/watch?v=zIX9IAunu_E

OECD. (2011). The Contribution of Agriculture to Green Growth. Pennsylvania State University, Agricultural and Environmental Economics. Retrieved from http://www.oecd.org/greengrowth/sustainable-agriculture/48258861.pdf

Our World in Data. (2018). Life Expectancy. Retrieved from https://ourworldindata.org/life-expectancy

Philips. (2015). How the Internet of Things is revolutionizing. Retrieved from https://www.philips.com/a-w/innovationmatters/blog/how-the-internet-of-things-isrevolutionizing-healthcare.html

Planet. (2018). Using space to help life on Earth. Retrieved from Monitor Earth, discover trends, deliver insights: https://www.planet.com/company/

Poitras, L. (Director). (2014). Citizenfour [Motion Picture]. United States, Germany: Radius-TWC.

Popular Science. (2015, April 20). To Predict Future Diseases, Doctors Will Map Newborns' Genes. Retrieved from http://www.popsci.com/doctors-will-map-newborns-genes-test-diseases

Powell, R. R. (2004). Wabi Sabi Simple. Adams Media.

Prisco, G. (2015, November 5). Slock.it to Introduce Smart Locks Linked to Smart Ethereum Contracts, Decentralize the Sharing Economy. Bitcoin Magazine. Retrieved from https://bitcoinmagazine.com/articles/slock-it-to-introduce-smart-locks-linked-to-smart-ethereum-contracts-decentralize-the-sharing-economy-1446746719/

Purves, D. (2007). Principles of Cognitive Neuroscience. Sinauer Associates Inc.

PwC. (2015). The Sharing Economy. Retrieved from https://www.pwc.com/us/en/technology/publications/assets/pwc-consumer-intelligence-series-the-sharing-economy.pdf

PwC. (2016). What mightblockchain mean for the mortgage industry. PwC US Fianancial Services. Retrieved from https://www.pwc.com/us/en/financial-services/publications/assets/pwc-financial-services-qa-blockchain-in-mortgage.pdf

Quartz. (2017, May 15). Your guide to understanding OBOR, China's new Silk Road plan. Retrieved from https://qz.com/983460/obor-an-extremely-simple-guide-to-understanding-chinas-one-belt-one-road-forum-for-its-new-silk-road/

Rauh, J. D. (2017). Hidden Debt, Hidden Deficits: 2017 Edition. New York: Hoover Institution. Retrieved from https://www.hoover.org/sites/default/files/research/docs/rauh_hiddendebt2017_final_webready1pdf1.pdf

Recruiting.com. (2018). Recruiting Strategy. Retrieved from The truth about millenial turnover: https://www.recruiting.com/blog/the-truth-about-millennial-turnover

ReferralMD. (2017). 30 Facts & Statistics On Social Media And Healthcare. Retrieved from https://getreferralmd.com/2017/01/30-facts-statistics-on-social-media-and-healthcare/

Reuters. (2016, December 22). U.N. warns of water crisis in Nigeria's megacity Lagos. Retrieved from https://www.reuters.com/article/us-un-nigeria-water/u-n-warns-of-water-crisis-in-nigerias-megacity-lagos-idUSKBN14B1V9

Reuters. (2017, September 18). Credit Suisse looks to make an impact with new investment wing. Retrieved from https://www.reuters.com/article/us-credit-suisse-gp-impactinvesting/credit-suisse-looks-to-make-an-impact-with-new-investment-wing-idUSKCN1BT15Z

Reuters. (2017, April 18). mHealth Market Worth $23 Billion in 2017 and Estimated to Grow at a CAGR of more than 35% over the next three years. Retrieved from https://www.reuters.com/brandfeatures/venture-capital/article?id=4640

Rifkin, J. (2015). Jeremy Rifkin: 'Number two cause of global warming emissions? Animal husbandry'. Euractive. Retrieved from https://www.euractiv.com/section/agriculture-food/interview/jeremy-rifkin-number-two-cause-of-global-warming-emissions-animal-husbandry/

Rifkin, J. (2015). Jeremy Rifkin: 'What's missing from UN climate talks is a new economic vision'. Euractiv. Retrieved from https://www.euractiv.com/section/sustainable-dev/interview/jeremy-rifkin-what-s-missing-from-un-climate-talks-is-a-new-economic-vision/

Rifkin, J. (2015). The zero marginal cost society. Jeremy Rifkin Enterprises.

Rifkin, J. (2017, January 17). The Third Industrial Revolution and a Zero Marginal Cost Society (Jeremy Rifkin) | DLD16. Retrieved from https://www.youtube.com/watch?v=5mQj574Cv_k

Roser, M. (2018). Our World in Data. Retrieved from War and Peace: https://ourworldindata.org/war-and-peace

Salvador Lopez, G. (2016). Strategy for the South European energy sector for the next 40 years. Stockholm: KTH Royal Institute of Technology.

Samuels, G. (2016). China carrying out over 60,000 illegal organ transplants annually, report finds. Independent. Retrieved from http://www.independent.co.uk/news/world/asia/china-carrying-out-millions-of-illegal-organ-transplants-annually-report-finds-a7107091.html

Santander InnoVentures, Oliver Wyman and Anthemis Group. (2015). The Fintech 2.0 Paper:rebooting financial services. Santander InnoVentures. Retrieved from http://santanderinnoventures.com/wp-content/uploads/2015/06/The-Fintech-2-0-Paper.pdf

Say no to palm oil. (2017). What's the issue? Retrieved from http://www.saynotopalmoil.com/Whats_the_issue.php

Scientific American. (1971, July). Next we will deliver people to live in it.

Scientific American. (2013, December). Sext much? If so, you're not alone. Retrieved from https://www.scientificamerican.com/article/sext-much-if-so-youre-not-alone/

Security Affairs. (2012). 7 Most Common Facebook Crimes. Retrieved from http://securityaffairs.co/wordpress/4891/cyber-crime/7-most-common-facebook-crimes.html

Siemens. (2013, August). Intelligent control centertechnology – Spectrum Power™. Retrieved from http://w3.siemens.com/smartgrid/global/SiteCollectionDocuments/Products_systems_solutions/Control%20Center/Spectrum%20Power%20Portfolio.pdf

Siemens. (2014). Scenario 2030: Riding to Reality. Pictures of the future, Digital Factory. Retrieved from https://www.siemens.com/innovation/en/home/pictures-of-the-future/industry-and-automation/digital-factories-riding-to-reality.html

Siemens. (2016). Do-it-all Drones. Pictures of the Future. Retrieved from https://www.siemens.com/innovation/en/home/pictures-of-the-future/digitalization-and-software/from-big-data-to-smart-data-smart-drones.html

Siemens. (2017, January 15). Retrieved from Offshore Wind Power - towards a sea change: http://www.siemens.com/global/en/home/markets/wind/offshore.html

Sinek, S. (2011, August 3). The Purpose Driven Organization w/ Simon Sinek. (O. Text, Ed.) Retrieved March 1, 2018, from https://www.youtube.com/watch?v=MXNvfEm5ezl

Sinek, S. (2016, October 29). Millennials in the Workplace. Retrieved from https://www.youtube.com/watch?v=hER0Qp6QJNU

SingularityHub. (2016, June 27). Why the World Is Better Than You Think in 10 Powerful Charts. Retrieved from https://singularityhub.com/2016/06/27/why-the-world-is-better-than-you-think-in-10-powerful-charts/#sm.00016theeftl9fr3zb91obzjxexie

Skinner, C. (2016, March 15). The five major use cases for financial blockchains. FinYear. Retrieved from http://www.finyear.com/The-five-major-use-cases-for-financial-blockchains_a35655.html

Solarly. (2016). Solarly presentation. Mont-Saint-Guibert: Internal publication. Retrieved from http://www.fichier-pdf.fr/2016/06/06/solarly-presentation/solarly-presentation.pdf

Stanford Encyclopedia of Philosophy. (2012, July 2). Entry: World Government. Retrieved from https://plato.stanford.edu/entries/world-government/

Statista. (2018). Retrieved from Number of available applications in the Google Play Store from December 2009 to December 2017: https://www.statista.com/statistics/266210/number-of-available-applications-in-the-google-play-store/

Stiglitz, J. E. (2002). Globalization and Its Discontents. New York: W.W. Norton & Company.

Strategy+business. (2017, September 27). Disruptors and the Disrupted: A Tale of Eight Companies – in Pictures. Retrieved from https://www.strategy-business.com/pictures/Disruptors-and-the-Disrupted-A-Tale-of-Eight-Companies-in-Pictures?gko=dd95b

TechCrunch. (2013, April 15). Asthmapolis Wants To Hack The Inhaler And Help 26 Million Americans Better Track And Manage Their Asthma. Retrieved from https://techcrunch.com/2013/04/05/asthmapolis-wants-to-hack-the-inhaler-and-help-26-million-americans-better-track-and-manage-their-asthma/

TechCrunch. (2014, October 11). Edward Snowden's Privacy Tips: "Get Rid Of Dropbox, Avoid Facebook And Google". Retrieved from https://techcrunch.com/2014/10/11/edward-snowden-new-yorker-festival/

TechCrunch. (2015, December 31). How To Stem The Global Shortage Of Data Scientists. Retrieved from https://techcrunch.com/2015/12/31/how-to-stem-the-global-shortage-of-data-scientists/

TechCrunch. (2017, March 8). Google confirms its acquisition of data science community Kaggle. Retrieved from https://techcrunch.com/2017/03/08/google-confirms-its-acquisition-of-data-science-community-kaggle/

Technavio. (2016, December 7). Is the building automation system the future of sustainable building maintenance? Retrieved from http://www.technavio.com/blog/building-automation-system-future-sustainable-building-maintenance

Techradar. (2016, May 30). 5 ways wearables will transform the lives of the elderly. Retrieved from https://www.techradar.com/news/wearables/5-ways-wearables-will-transform-the-lives-of-the-elderly-1321898

TED. (2012, February). Connected, but alone? Retrieved from https://www.ted.com/talks/sherry_turkle_alone_together

TED (Producer). (2014). Think your e-mail is private? Think again. [Motion Picture]. Retrieved from https://www.ted.com/talks/andy_yen_think_your_email_s_private_think_again/discussion?language=en

Telegraph. (2010, October 20). Ten year olds have 7000 worth of toys but play with just 330. Retrieved from http://www.telegraph.co.uk/finance/newsbysector/retailandconsumer/8074156/Ten-year-olds-have-7000-worth-of-toys-but-play-with-just-330.html

Terry Hartig, P. H. (2016, May 20). Living in cities, naturally. Science, pp. Vol. 352, Issue 6288. Retrieved from http://science.sciencemag.org/content/352/6288/938.full, http://www.futurity.org/cities-nature-1178592-2/

Than, K. (2016). Larger marine animals at higher risk of extinction, and humans are to blame, Stanford-led study finds. Stanford University, Stanford School of Earth, Energy & Environmental Sciences. Retrieved from http://news.stanford.edu/2016/09/14/larger-marine-animals-higher-risk-extinction-humans-blame/

The Children Society. (2018). Safety Net: Cyberbullying's impact on young people's mental health. Young Minds. Retrieved from https://www.childrenssociety.org.uk/sites/default/files/social-media-cyberbullying-inquiry-full-report_0.pdf

The Economist. (2015). Cool, man. London: The Economist Newspaper Limited. Retrieved from http://www.economist.com/news/special-report/21650291-where-small-businesses-can-borrow-if-banks-turn-them-down-cool-man

The Economist. (2016, June 11). Feeding the ten billion. The Economist.

The Economist. (2016). Technology Quaterly: The future of agriculture. Technology Quaterly. Retrieved from http://www.economist.com/technology-quarterly/2016-06-09/factory-fresh

The Economist. (2018, April 19). What's in it for the Belt-and-Road countries? Retrieved from https://www.economist.com/the-economist-explains/2018/04/19/whats-in-it-for-the-belt-and-road-countries

The Guardian. (2006, November 20). Nike sacks Pakistani supplier over child labour row. Retrieved from https://www.theguardian.com/business/2006/nov/20/2

The Guardian. (2014, May 5). Free lunch, anyone? Foodsharing sites and apps stop leftovers going to waste. Retrieved from https://www.theguardian.com/sustainable-business/free-food-sharing-leftovers-surplus-local-popular

The Guardian. (2014, November 2). How the World Was Won: The Americanization of Everywhere review – a brilliant essay. Retrieved from https://www.theguardian.com/books/2014/nov/02/how-the-world-was-won-americanization-of-everywhere-review-peter-conrad

The Guardian. (2015). Global population set to hit 9.7 billion people by 2050 despite fall in fertility. Retrieved from https://www.theguardian.com/global-development/2015/jul/29/un-world-population-prospects-the-2015-revision-9-7-billion-2050-fertility

The Guardian. (2015, July 8). NSA tapped German Chancellery for decades, WikiLeaks claims. Retrieved from https://www.theguardian.com/us-news/2015/jul/08/nsa-tapped-german-chancellery-decades-wikileaks-claims-merkel

The Guardian. (2016, August 31). Dropbox hack leads to leaking of 68m user passwords on the internet. Retrieved from https://www.theguardian.com/technology/2016/aug/31/dropbox-hack-passwords-68m-data-breach

The Guardian. (2016, June 22). Mark Zuckerberg tapes over his webcam. Should you? Retrieved from https://www.theguardian.com/technology/2016/jun/22/mark-zuckerberg-tape-webcam-microphone-facebook

The Guardian. (2017, June 18). Life and death in Apple's forbidden city. Retrieved from https://www.theguardian.com/technology/2017/jun/18/foxconn-life-death-forbidden-city-longhua-suicide-apple-iphone-brian-merchant-one-device-extract

The Guardian. (2017, July 21). Pepsico, Unilever and Nestlé accused of complicity in illegal rainforest destruction. Retrieved from https://www.theguardian.com/environment/2017/jul/21/pepsico-unilever-and-nestle-accused-of-complicity-in-illegal-rainforest-destruction

The Guardian. (2017, May 8). The meaning of life in a world without work. Retrieved from https://www.theguardian.com/technology/2017/may/08/virtual-reality-religion-robots-sapiens-book

The Guardian. (2017, June 26). What jobs will still be around in 20 years? Read this to prepare your future. Retrieved from https://www.theguardian.com/us-news/2017/jun/26/jobs-future-automation-robots-skills-creative-health

The Huffington Post. (2011). Credit Rating Agency Analysts Covering AIG, Lehman Brothers Never Disciplined. Retrieved from http://www.huffingtonpost.com/2009/09/30/credit-rating-agency-anal_n_305587.html

The Nature Conservancy. (2018, January). Conserving the lands and waters on which all life depends. Retrieved from https://www.nature.org/?intc=nature.tnav.logo

The State Council of the People's Republic of China. (2015, March 28). China unveils action plan on Belt and Road Initiative. Xinhua News Agency. Retrieved from http://english.gov.cn/news/top_news/2015/03/28/content_281475079055789.htm

The Telegraph. (2009, June 30). Genetic 'MoT' for disease free babies. Retrieved from https://www.telegraph.co.uk/news/health/news/5702674/Genetic-MoT-for-disease-free-babies.html

The Telegraph. (2014, May 21). More than 15 million Britons at risk of identity theft after eBay hacked. Retrieved from https://www.telegraph.co.uk/technology/news/10848080/More-than-15-million-Britons-at-risk-of-identity-theft-after-eBay-hacked.html

The Wall Street Journal. (2011, April 23). Number of the Week: Americans Buy More Stuff They Don't Need. Retrieved from https://blogs.wsj.com/economics/2011/04/23/number-of-the-week-americans-buy-more-stuff-they-dont-need/

The Wall Street Journal. (2016, August 22). Retrieved from ChemChina-Syngenta $43 Billion Deal Approved by U.S. Security Panel: http://www.wsj.com/articles/u-s-security-watchdog-clears-43-billion-chemchina-syngenta-takeover-deal-1471844896

The Washington Post. (2016, December 30). China's $9 billion effort to beat the U.S. in genetic testing. Retrieved from https://www.washingtonpost.com/news/wonk/wp/2016/12/30/chinas-9-billion-effort-to-beat-the-u-s-in-genetic-testing/?utm_term=.d5217639b49b

The World Factbook. (2015). Country Comparison to the World of Literacy Rate. Central Intelligence Agency. Retrieved from https://www.cia.gov/library/publications/the-world-factbook/fields/2103.html#136

Time. (2013, May 20). Millennials: The Me Me Me Generation. Time. Retrieved from http://time.com/247/millennials-the-me-me-me-generation/

Time. (2015, March 18). You Asked: How Many Friends Do I Need? Time. Retrieved from http://time.com/3748090/friends-social-health/?iid=time_speed

Time. (2018, May 1). Scientists Announce Plan to Create Virus-Proof Cells. Retrieved from http://time.com/5261777/scientists-virus-proof-cells/

UBS. (2016). Extreme automation and connectivity: The global, regional, and investment implications of the Fourth Industrial Revolution. UBS White Paper for the World Economic Forum. Retrieved from https://www.ubs.com/global/en/about_ubs/follow_ubs/highlights/davos-2016.html

UNESCO. (2015). Adult literacy rate, population 15+ years (both sexes, female, male). Retrieved from http://data.uis.unesco.org/Index.aspx?DataSetCode=EDULIT_DS&popupcustomise=true&lang=en#

Unite. (2015). New ways of working. The B team. Virgin Unite. Retrieved from https://issuu.com/the-bteam/docs/150114_newwaysofworking_v12?e=15214291/11024330

United Nations. (2006). Rearing cattle produces more greenhouse gases than driving cars, UN report warns. Retrieved from https://news.un.org/en/story/2006/11/201222-rearing-cattle-produces-more-greenhouse-gases-driving-cars-un-report-warns

United Nations. (2016). Sustainable Development Goals. United Nations Development Programme. United Nations. Retrieved from http://www.undp.org/content/undp/en/home/sustainable-development-goals/

United Nations. (2016). The World's Cities in 2016. Retrieved from http://www.un.org/en/development/desa/population/publications/pdf/urbanization/the_worlds_cities_in_2016_data_booklet.pdf

United Nations. (2017, June 21). World population projected to reach 9.8 billion in 2050, and 11.2 billion in 2100. Retrieved from https://www.un.org/development/desa/en/news/population/world-population-prospects-2017.html

United Nations. (2018). Retrieved March 3, 2018, from Climate Change: http://www.un.org/en/sections/issues-depth/climate-change/

United Nations. (2018, February). Paris Agreement - Status of Ratification. Retrieved from United Nations - Climate Change: http://unfccc.int/paris_agreement/items/9444.php

United States Evironmental Protection Agency. (2018, March). Understanding Global Warming Potentials. Retrieved from https://www.epa.gov/ghgemissions/understanding-global-warming-potentials

Upwork. (2015, December 14). The 4 Trends That Will Change the Way We Work by 2021. Retrieved from https://www.upwork.com/hiring/trends/four-trends-will-change-way-work-2021/

US Energy Information Administration. (2016). Frequently asked question. Retrieved December 24, 2016, from https://www.eia.gov/tools/faqs/faq.cfm?id=85&t=1

Valve. (2012). Handbook for new employees. Retrieved from http://www.valvesoftware.com/company/Valve_Handbook_LowRes.pdf

Vcloud News. (2015, April 5). EVERY DAY BIG DATA STATISTICS - 2.5 QUINTILLION BYTES OF DATA CREATED DAILY. Retrieved from http://www.vcloudnews.com/every-day-big-data-statistics-2-5-quintillion-bytes-of-data-created-daily/

Verhelst, J.-L. (2017). Bitcoin the blockchain and beyond. Brussels.

Vox. (2017, November 13). The big debate about the future of work, explained. Retrieved from https://www.youtube.com/watch?v=TUmyygCMMGA

Vox. (2017, December 12). The diet that helps fight climate change. Retrieved from https://www.youtube.com/watch?v=nUnJQWO4YJY

Walla, K. (2006, November). Fact and Fiction in Organ Transplant Waiting List Fraud. Retrieved from https://www.law.uh.edu/healthlaw/perspectives/2006/(KW)TransplantFraud.pdf

Water footprint. (2005). Personal water footprint calculator. Retrieved from http://waterfootprint.org/en/resources/interactive-tools/personal-water-footprint-calculator/

WebMD. (2017, January 15). About WebMD Policies. Retrieved from What We Do For Our Users: http://www.webmd.com/

Wen, T. (2014, November 10). Why Don't More People Want to Donate Their Organs? The Atlantic. Retrieved from https://www.theatlantic.com/health/archive/2014/11/why-dont-people-want-to-donate-their-organs/382297/

Wikipedia. (2018). Retrieved July 8, 2018, from Definition of remittance: https://en.wikipedia.org/wiki/Remittance

Wikipedia. (2018). Deepwater Horizon oil spill. Retrieved March 3, 2018, from https://en.wikipedia.org/wiki/Deepwater_Horizon_oil_spill

Wind Europe Association. (2016, February 15). Retrieved from The European offshore wind industry - key trends and statistics 2015: https://windeurope.org/about-wind/statistics/offshore/key-trends-2015/

Wired. (2017, November 5). The US government isn't just tech-illeterate, it's tech-incompetent. Retrieved from https://www.wired.com/2017/05/real-threat-government-tech-illiteracy/

Wired. (2018). The Wired guide to CRISPR. California. Retrieved from https://www.wired.com/story/wired-guide-to-crispr/

World Bank. (2016, November 4). Seeking Solutions to Fill the Savings Gap. Retrieved from http://www.worldbank.org/en/news/feature/2016/11/04/seeking-solutions-to-fill-the-savings-gap

World Bank. (2016, September 20). The world's top 100 economies: 31 countries; 69 corporations. World
    Bank Blog. Retrieved from http://blogs.worldbank.org/publicsphere/world-s-top-100-
    economies-31-countries-69-corporations

World Bank. (2016). Toolkit on Intelligent Transport Systems for Urban Transport. Retrieved from Fuel
    consumption monitoring:
    https://www.ssatp.org/sites/ssatp/files/publications/Toolkits/ITS%20Toolkit%20content/its-
    applications/its-facilitated-functions/fuel-consumption-monitoring.html

World Bank. (2017). Remittances to Developing Countries Decline for Second Consecutive Year.
    Washington. Retrieved from http://www.worldbank.org/en/news/press-
    release/2017/04/21/remittances-to-developing-countries-decline-for-second-consecutive-
    year

World Economic Forum. (2011). Personal Data:The Emergence of a New Asset Class. Retrieved from
    http://www3.weforum.org/docs/WEF_ITTC_PersonalDataNewAsset_Report_2011.pdf

World Economic Forum. (2014). A Framework for Sustainable Security Systems. Global Agenda Council
    on Social Security Systems. Retrieved from
    http://www3.weforum.org/docs/GAC/2014/WEF_GAC_SocialSecuritySystems_SustainableSe
    curitySytems_Framework_2014.pdf

World Economic Forum. (2015). Intelligent Assets Unlocking the Circular Economy Potential. Industrial
    agenda, Davos. Retrieved from
    http://www3.weforum.org/docs/WEF_Intelligent_Assets_Unlocking_the_Cricular_Economy.p
    df

World Economic Forum. (2016). Chapter 1: The Future of Jobs and Skills. Davos. Retrieved from
    http://reports.weforum.org/future-of-jobs-2016/chapter-1-the-future-of-jobs-and-
    skills/#view/fn-1

World Health Organization. (2018). Sustainable Cities Health at the Heart of Urban Development.
    Retrieved from Health and sustainable development goals: http://www.who.int/sustainable-
    development/cities/Factsheet-Cities-sustainable-health.pdf?ua=1

Zacks Equity Research. (2015, October 20). Yahoo! Finance. Retrieved from Is SolarCity Changing the
    World of Solar Power?: http://finance.yahoo.com/news/solarcity-changing-world-solar-
    power-224510495.html

ABOUT
THE AUTHOR

Gerard Salvador López

I am that guy that is in love with technology, strategy and economy.

Think global, be thankful and always stay humble.

We are living an exciting moment in history!

For more information:
www.gerardsl.com

www.linkedin.com/in
/gsalvador

For any question, request
or whatever:
hi@gerardsl.com

www.ingramcontent.com/pod-product-compliance
Lightning Source LLC
Chambersburg PA
CBHW032329210326
41518CB00041B/1922